Symmetry

Symmetrie: Hans Walser second edition
© B. G. Teubner, Stuttgart, 1996.
Translation from the original German edition arranged
with the approval of the publisher, B. G. Teubner.

© 2000 by
The Mathematical Association of America (Incorporated)
Library of Congress Catalog Card Number 00-107972

ISBN 0-88385-532-1

Printed in the United States of America

Current Printing (last digit):
10 9 8 7 6 5 4 3 2 1

Symmetry

Hans Walser

Translated from the original German by Peter Hilton,
with the assistance of Jean Pedersen

Published and Distributed by
THE MATHEMATICAL ASSOCIATION OF AMERICA

SPECTRUM SERIES

The Spectrum Series of the Mathematical Association of America was so named to reflect its purpose: to publish a broad range of books including biographies, accessible expositions of old or new mathematical ideas, reprints and revisions of excellent out-of-print books, popular works, and other monographs of high interest that will appeal to a broad range of readers, including students and teachers of mathematics, mathematical amateurs, and researchers.

All the Math That's Fit to Print, by Keith Devlin
Circles: A Mathematical View, by Dan Pedoe
Complex Numbers and Geometry, by Liang-shin Hahn
Cryptology, by Albrecht Beutelspacher
Five Hundred Mathematical Challenges, Edward J. Barbeau, Murray S. Klamkin, and William O. J. Moser
From Zero to Infinity, by Constance Reid
I Want to be a Mathematician, by Paul R. Halmos
Journey into Geometries, by Marta Sved
JULIA: a life in mathematics, by Constance Reid
The Lighter Side of Mathematics: Proceedings of the Eugène Strens Memorial Conference on Recreational Mathematics & its History, edited by Richard K. Guy and Robert E. Woodrow
Lure of the Integers, by Joe Roberts
Magic Tricks, Card Shuffling, and Dynamic Computer Memories: The Mathematics of the Perfect Shuffle, by S. Brent Morris
The Math Chat Book, by Frank Morgan
Mathematical Carnival, by Martin Gardner
Mathematical Circus, by Martin Gardner
Mathematical Cranks, by Underwood Dudley
Mathematical Fallacies, Flaws, and Flimflam, by Edward J. Barbeau
Mathematical Magic Show, by Martin Gardner
Mathematics: Queen and Servant of Science, by E. T. Bell
Memorabilia Mathematica, by Robert Edouard Moritz
New Mathematical Diversions, by Martin Gardner
Non-Euclidean Geometry, by H. S. M. Coxeter
Numerical Methods that Work, by Forman Acton
Numerology or What Pythagoras Wrought, by Underwood Dudley

MAA Service Center
P.O. Box 91112
Washington, DC 20090-1112
800-331-1622   FAX 301-206-9789

Foreword to the German Edition

The idea of *symmetry* can be viewed in very different ways: The narrowest interpretation is limited to two-sided symmetry, as applied, more or less exactly, to the external form of the human body. The broadest interpretation understands by symmetry the property of anything that is in some way regular and shows repetitions. Thus the cycle of seasons is symmetrical, since it repeats itself periodically; but so, too, is a four-stroke engine, the decimal expansion of $\frac{1}{7}$, a carpet pattern, an ornament or even a song or a poem. In this sense we meet symmetry practically everywhere, especially in science and art. Moreover, modern engineering production methods lead to highly similar and in this sense mutually symmetric products.

The object of this book is to present selected examples of symmetry in an understandable way. We do not aim for systematic completeness, but the readers are given references to literature that will lead them further. For me it is most important to "sharpen the eye" for the proper perception of symmetry in the world around us. It will also be shown how symmetry can be used as a methodological working tool.

On the theme of symmetry there is an extensive literature, often concentrating on a particular aspect. A classic, from any point of view, is the book by Hermann Weyl [52]. In [20] physical and chemical aspects of symmetry are predominantly treated; [45] links this up with philosophical reflections. Aspects of symmetry in 2- and 3-dimensional geometry and in engineering are described in [40]. Symmetry also plays a central role in various areas of art, particularly ornamental art (compare [2], [7], [14]). One should especially mention in this context the graphics of Maurits Cornelis Escher (compare [13], [32]). Finally, symmetry has repeatedly been the theme of exhibitions and conferences (compare [36], [51]).

This book is directed towards students, schoolgirls and schoolboys, as well as to their teachers and interested laymen. The text is modular in construction, so that the individual chapters may be read independently of each other. Questions are distributed throughout the text; these relate in part to further aspects of the subject. The answers to these questions are gathered together at the end of each chapter.

Many of the ideas worked out in this book go back to suggestions by my students, to whom I am most grateful. For some examples I must thank my colleague Peter Gallin. I owe particular thanks to my colleague Reto Schuppli for a critical review of the text. I thank Mr. Jürgen Weiss of B. G. Teubner, Leipzig publishers, for his generous supervision of this work.

Frauenfeld, December 1997   Hans Walser

Foreword to the English Edition

We have been faithful to the original German text, except that we have not attempted the impossible task of translating German poetry or poetry in Swiss dialect into appropriate English—we have simply omitted it. This reasonable, if somewhat unambitious, course has the author's full approval, but it has, of course, led to a rather thin Chapter 6.

In one respect our fidelity to the original has unfortunate consequences for the reader. We have retained the original references to German language sources, even where an English language version of the reference exists— for example, the reference to the classic text by Hermann Weyl in the author's foreword. It has only rarely been possible to find English language alternatives to the original references, since the author has always chosen his references to make, or to reinforce, very specific points. Where an appropriate English language reference has been found it has been inserted but the German reference has not been omitted. This accounts for the innovations in the enumeration of the references.

We must emphasize that this monograph is not a comprehensive text covering an area of geometry. It was written to attract the reader to geometrical ideas, especially those related to the concept of symmetry.

It is a pleasure to acknowledge the invaluable, and highly efficient, assistance of my colleague Jean Pedersen in making this translation available to the English-speaking—and non-German-speaking—world; and to express my appreciation to my colleague Jerry Alexanderson for his careful perusal of the translation. Finally, I am happy to acknowledge the invaluable assistance of my colleagues Rudolf Fritsch and Branko Grünbaum in greatly improving an early version of the translation.

Binghamton, June 2000   Peter Hilton

Author's Note to English Addition

It is a pleasure to express my appreciation of the careful work done by my colleagues Peter Hilton and Jean Pedersen in making available an English version of my text.

I would also like to take this opportunity to thank Jerry Alexanderson for his editorial work, and Elaine Pedreira Sullivan and Beverly Ruedi for their careful attention to detail in the production of this translation.

Hans Walser

Contents

CHAPTER 1

Little Mirror, Little Mirror

1.1 EVER FURTHER INWARDS

1.1.1 The Mirror in a Mirror in a Mirror

If we hold a round pocket-mirror close to our eye and stare into the bathroom mirror, we see, by parallel placement of the two mirrors, a sequence of ever smaller circles as mirror images of the pocket-mirror (Figure 1). How are the radii of these circles related to each other? (It's worthwhile actually conducting the experiment and estimating the diameters of the circles.) Let

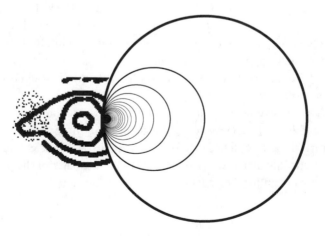

FIGURE 1
The mirror in a mirror

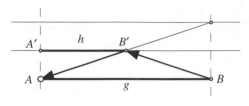

FIGURE 2
The first image of the pocket mirror

us study the situation with the help of a diagram. The segment $\overline{AB}$ on the straight line g represents the pocket mirror, looked at from above, and the straight line h represents the parallel bathroom mirror. Suppose the observing eye is at the point A. Then we obtain our first mirror image of the pocket mirror according to Figure 2.

We mirror the endpoint B of the pocket mirror in the line h and join the point so obtained to the eye A. The intersection B' with the line h marks the place where the point B will be seen in the bathroom mirror. The diameter $\overline{A'B'}$ of the first image of the pocket-mirror is thus half the length of the real diameter $\overline{AB}$ of the pocket-mirror.

Question 1.1 How high must a wall-mirror be, so that one can see oneself in it from the top of one's head to the soles of one's feet?

The second (next smaller) image of the pocket mirror arises by reflecting the line of sight one additional time, backwards and forwards (Figure 3).

The image-diameter $\overline{A''B''}$ is now just a quarter of the true diameter $\overline{AB}$. At this point I conjectured that, at the next stage, an eighth of the original diameter would appear. This conjecture, however, is wrong; Figure 4 shows that the third image of the pocket mirror has a diameter which is one sixth of the true diameter.

Each successive image requires an additional reflection in the bathroom mirror and in the pocket mirror itself; the diameter of the nth image is the $\frac{1}{2n}$th part of the true diameter of the pocket mirror. If, instead of the true pocket mirror diameter, we take one half of the diameter of the first image as unit, we obtain for the image-diameters the sequence

$$1, \frac{1}{2}, \frac{1}{3}, \frac{1}{4}, \frac{1}{5}, \cdots$$

This sequence is called the ***harmonic sequence.***

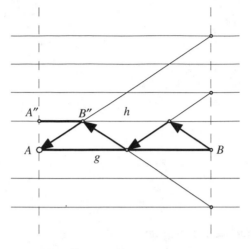

FIGURE 3
The second image of the pocket mirror

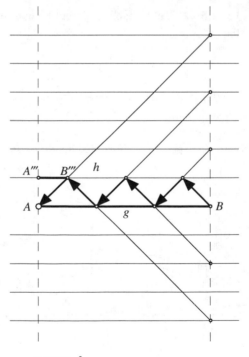

FIGURE 4
The third image of the pocket mirror

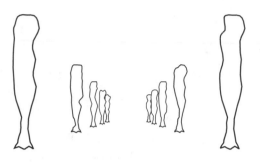

FIGURE 5
Avenue of poplars

1.1.2 An Avenue of Poplars

Such a harmonic sequence also appears in the theory of perspectives. Figure 5 shows, as an example of this, an avenue of poplars where, in fact, the poplars have the same height and are equally spaced, and the observer is next to a tree.

The poplars have, from front to back, apparent heights shortened by the perspective, in the ratios

$$1 : \frac{1}{2} : \frac{1}{3} : \frac{1}{4} : \frac{1}{5} \cdots$$

This can be seen in Figure 6, where the view from the side is given. The eye of the observer is at the point A; the perspectively shortened height $\overline{B'C'}$ of the second tree is then one half of the height $\overline{BC}$ of the first tree. For the third tree we obtain a perspectively shortened height of one third, etc.

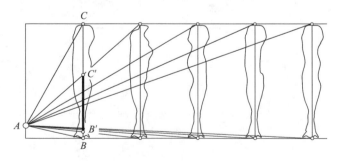

FIGURE 6
View from the side

FIGURE 7
The monitor in the monitor in the monitor

1.1.3 The Monitor in the Monitor

There is, however another "picture in the picture" situation, in which it is not the harmonic sequence but a geometric sequence which turns up as the sequence of length-ratios: If we place a video camera in front of a monitor and show the image on the monitor, we obtain the situation in Figure 7.

We see—at least in theory—infinitely many monitors in the monitor. In practice, in carrying out this experiment, there soon occur problems with the intensity of the light, so that only a few monitors, boxed inside each other, can be recognized.

If now the length of the second monitor is reduced by a factor $f, 0 < f < 1$, with respect to the first, the next monitor undergoes a shortening factor of f^2. The lengths of the monitors, boxed one inside the other, are thus related as in the geometric sequence

$$1 : f : f^2 : f^3 : f^4 \cdots$$

A self-similarity occurs here: a partial picture, which consists of an arbitrarily chosen image and all subsequent images, is similar, in the sense of a central contraction, to the entire picture, the contraction factor (ratio of magnification) is a power of f, and the center of contraction is the point of accumulation "at infinity".

FIGURE 8
Four monitors

If we project back several monitors with a video camera, for example four monitors, as in Figure 8, there occurs a situation, where there is not just one, unique self-similarity center, but infinitely many.

In this situation we speak of a *fractal* [34]. In Figure 9 one of the monitors is defective, and this has pervasive consequences, in that the dark monitor reappears in infinitely many places in the other three monitors.

Figure 10 shows the situation in schematic form for the original monitors and the first generation of images.

Figure 11 finally shows also all the subsequent generations, thus the complete fractal.

FIGURE 9
A monitor is defective

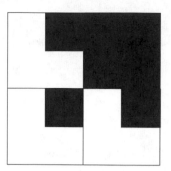

FIGURE 10
Schema

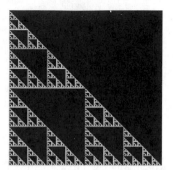

FIGURE 11
The Fractal

Question 1.2 How big is the black part of Figure 11?

Question 1.3 How long is the diagonal from left top to right bottom?

A black-white color inversion produces from Figure 11 an affinely distorted **Sierpinski Triangle** [34] (Figure 12).

1.2 SEEN FROM THE SIDE

In our discussion thus far, we were ourselves "in the picture"—we were drawn into the optical events. Now we will try to stay outside the line of sight, that is, to a certain extent to look at two mirrors from the side.

At the start of our considerations we have two parallel mirrors g and h. If some shape, for example the ship in position 0 of Figure 13, is reflected in g,

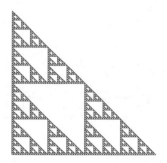

FIGURE 12
The Sierpinski Triangle

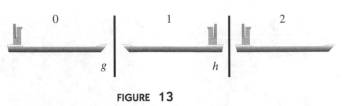

FIGURE **13**
Reflections in g and h

we obtain a mirror-image ship (position 1). This ship has a "false", that is, a mirror-image spatial orientation. For example, the red side-lantern appears, on this mirror-image ship, on the starboard side instead of the port side, and the propeller, normally a right-handed screw, would now be a left-handed screw.

Question 1.4 Does the ship now appear to be going backwards?

Through a further reflection, this time in h, we obtain the ship in position 2. Here the red side-lantern is again on the right (starboard) side. The direct passage from position 0 to position 2 is a ***translation.*** The ship could simply have traveled forward twice the distance from g to h. The doubled distance vector from g to h is then called the ***translation vector.*** If we reflect now in g, then in h, then a second time in g and a second time in h, we obtain position 4 (Figure 14), which corresponds to a translation by twice the given translation vector.

If we imagine this translation carried out infinitely often, there appears a ***translation-symmetric*** figure (Figure 15).

A translation-symmetric figure thus has no end and, since we may think of the translation as carried out backwards infinitely often, also no beginning. In practice it is thus always only a section of a translation-symmetric figure which can be represented.

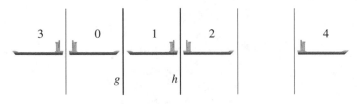

FIGURE **14**
Multiple reflection in parallel planes

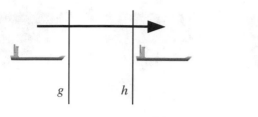

FIGURE 15
Translation symmetry

Answers to Questions

Answer 1.1 The mirror must be at least half as high as the person in question. Moreover, the mirror must be hung at the correct height: the lower edge must be at half the height of the eyes of the person being mirrored (Figure 16). Also one cannot then see the soles of the feet.

Answer 1.2 For the surface area of the black squares we have:

$$s = \frac{1}{4} + 3\left(\frac{1}{4}\right)^2 + 9\left(\frac{1}{4}\right)^3 + 27\left(\frac{1}{4}\right)^4 + \cdots$$

$$= \frac{1}{4}\left[1 + \frac{3}{4} + \left(\frac{3}{4}\right)^2 + \left(\frac{3}{4}\right)^3 + \cdots\right] = \frac{1}{4}\frac{1}{1-\frac{3}{4}} = 1$$

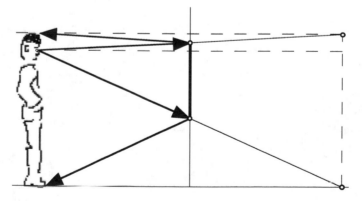

FIGURE 16
Wall-mirror

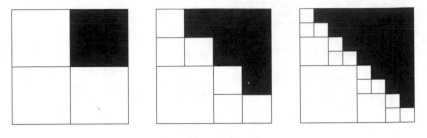

FIGURE 17
The boundary of steps

Thus everything appears to have become black. In the fractal of Figure 11 the design has been faked, namely drawn so that not all the generations, but only the first eight, are represented.

Answer 1.3 It follows from Pythagoras' Theorem that the length of the diagonal of a square is $\sqrt{2}$ times the length of the side of the square. But in Figure 11 the diagonal is the boundary of steps of ever finer height (Figure 17). Each step, however, vertical and horizontal counted together, is twice as long as a side of the square. Viewed in this way, the "diagonal", as boundary of such steps, also has this length.

Answer 1.4 The ship in mirror-image position 1 would, relative to its own bow, travel forwards, since the engine would also have changed the rotational direction of the shaft.

CHAPTER 2

Inside and Outside

2.1 REFLECTING IN A CIRCLE[1]

If, on a square grid, a gridline h is chosen as axis of a straight-line reflection, the reflection of lattice points can be done by "counting squares" (Figure 18). In this way one half-plane is mapped onto the other half-plane.

Now we try, with a comparable procedure, to map the interior of a circle on the exterior, and the other way round. This seems at first sight self-contradictory, since the outside of a circle is much "bigger" than its interior. However, Figure 19 shows a "square structure", in which the "squares" become smaller as one moves inwards. Thus, indeed, can a mapping of A onto A' be achieved, square by square.

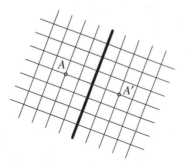

FIGURE 18
Reflection in a gridline on a square grid

---

[1] Here, and subsequently, the author has "reflection" where many standard English-language texts would speak of "inversion."

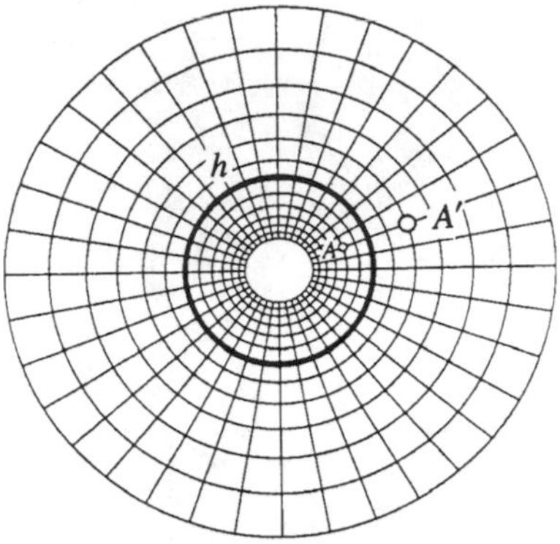

FIGURE 19
Reflection in a circle

This mapping we call a ***circle-reflection*** (or ***inversion***). The highlighted circle h is called the ***reference circle;*** it corresponds to the axis of a straightline-reflection. In constructing such a "circular square" pattern we have choices, so we must clarify how the radii of the concentric circles for the scaling $t = \cdots, -2, -1, 0, 1, 2, \cdots$ are related to each other (Figure 20). For that purpose, we imagine the "square pattern" further subdivided, so that the quadrilaterals are almost squares.

We now study such a "near-square", lying in a circular sector of angle α radians at a distance r from the center of the circle (Figure 21). This near-square has a side of length Δr in the radial direction and a side of length

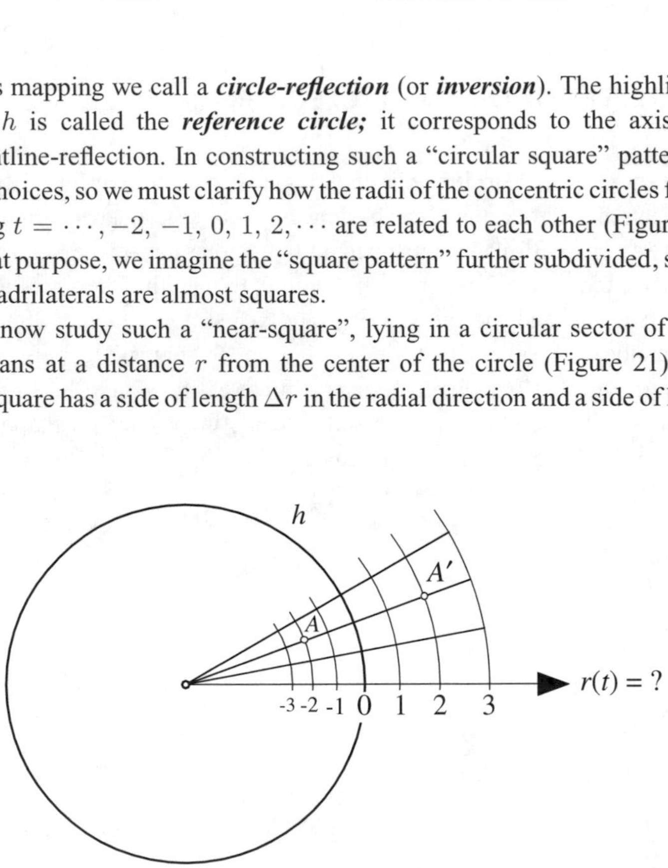

FIGURE 20
$r(t) = ?$

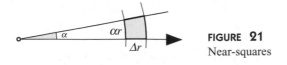

FIGURE **21**

Near-squares

αr in the tangential direction. We may reason heuristically as follows. These two lengths should be equal, so that $\Delta r = \alpha r$. In terms of differentials we obtain $dr = \alpha r$, so, through separation of variables,

$$\frac{dr}{r} = \alpha,$$

and thus, through integration, $\ln r = \alpha t + C_1$, so that, finally, $r(t) = Ce^{\alpha t}$. If we denote the radius of the reference-circle h by r_0, we get, from the boundary condition $r(0) = r_0$, the function

$$r(t) = r_0 e^{\alpha t}.$$

The radii depend exponentially on t. As $t \to -\infty$, $r \to 0$; thus the interior of the circle has room for infinitely many "squares". For the further investigation of the mapping of A onto A' polar coordinates are appropriate (Figure 22).

The point A and its image A' have the same polar angle ϕ, and their polar distances belong to equal and opposite t-values: thus

$$r = r_0 e^{\alpha t}, \quad r' = r_0 e^{-\alpha t}.$$

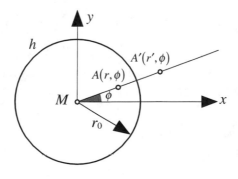

FIGURE **22**

Polar coordinates

Thus we obtain

$$rr' = r_0^2.$$

Remark If we choose a unit circle for the reference circle h, so that $r_0 = 1$, then the polar distances r and r' of a point and its image are reciprocals of each other. In this case the mapping can be described in the complex plane by $w = f(z) = \frac{1}{\bar{z}}$.

2.2 COMPOSITION OF TWO CIRCLE-REFLECTIONS

We now study the case of two circle-reflections (or inversions) whose reference-circles h and H are concentric, with radii r_0 and R_0. Under reflection in h the point A is mapped to the point A'; by reflection in H the point A' is mapped to the point A'' (Figure 23). In the passage from A to A' we have $rr' = r_0^2$; in going from A' to A'' we have $r'r'' = R_0^2$. For the passage from A to A'' we thus have $\frac{r''}{r} = \frac{R_0^2}{r_0^2}$; so the composition of two circle-reflections with concentric reference-circles is a central dilatation with magnification ratio $\frac{R_0^2}{r_0^2}$; the center of dilatation is the common center of the two reference-circles.

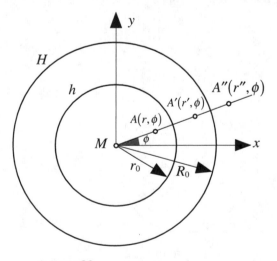

FIGURE 23
The composition of two circle-reflections

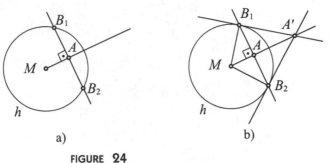

FIGURE 24
Direct construction of the image point

2.3 DIRECT CONSTRUCTION OF THE IMAGE POINT

Figure 24 shows in two steps how, for a given point A in the interior of the reference-circle h, the image point A' may be constructed.

To prove the correctness of this construction we consider the right-angled triangle $MA'B_1$ (Figure 24b) with its right angle at B_1. In this triangle r_0 is the length of MB_1, r is the length of the part MA of the hypotenuse, and r' is the length of the hypotenuse. By the similarity of the triangles MAB_1 and MB_1A' we have $rr' = r_0^2$.

Question 2.1 How may the point A' be constructed, when its pre-image A lies outside the reference-circle h?

Question 2.2 Which are the fixed points of a circle-reflection?

2.4 CIRCLE-REFLECTION INVARIANTS

The usual straight-line reflection is a congruence mapping: thus the images of straight lines, circles, squares, etc. are again straight lines, circles, squares, etc. The circle-reflection is not a congruence mapping, since the bounded interior of the reference-circle is mapped onto the unbounded exterior. Nevertheless some configurations are such that their images are of the same kind.

A bishop on a chessboard follows along the diagonals of neighboring ("by corners") chessboard squares. A sequence of such diagonals lies naturally on a straight line. The situation is very different in the case of the square grid of Figure 19. Here a sequence of square-diagonals gives rise to a curved

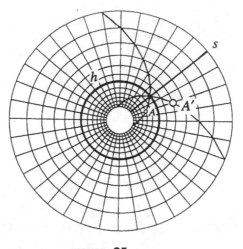

FIGURE 25
"Diagonal curves"

line of spiral shape (Figure 25). Precisely, a sequence of square-diagonals gives rise to a broken line with vertices at the grid points of the square grid. But if we subdivide the square grid ever finer and think of it provided with square-diagonals, there arises a beautifully curved spiral. Reflection of this spiral in the reference-circle h, which maps A to A', gives rise to a spiral of exactly the same kind. The two spirals are thus mirror images with respect to circle-reflection in h. Moreover, they are obviously also mirror images with respect to straight line reflection in the radial grid line s. For such a diagonal spiral the polar angle ϕ increases, for each grid step in the radial direction, by the same value α; α is the angle of intersection of two neighboring radial grid lines. If we denote by t the number of grid steps from h towards the exterior, then the relation $\phi = \alpha t$ holds, so that $t = \frac{\phi}{\alpha}$. Since $r(t) = r_0 e^{\alpha t}$, the diagonal spiral is represented in polar coordinates by the equation

$$r\left(\frac{\phi}{\alpha}\right) = r_0 e^{\phi}.$$

The diagonal spiral is thus the graph of an exponential function drawn in polar coordinates. In general a spiral with the polar representation

$$r(\phi) = r_0 e^{m\phi}$$

is called a ***logarithmic spiral,*** since the polar angle ϕ depends logarithmically on the polar distance r. Figure 26 shows a logarithmic spiral with $m = \frac{1}{5}$.

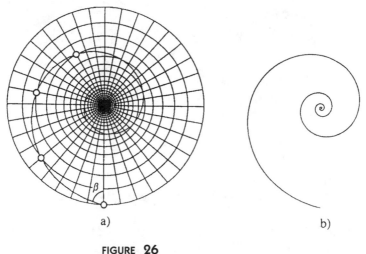

a)   b)

FIGURE 26

A logarithmic spiral with $m = \frac{1}{5}$

With this spiral a unit-square in the radial direction arises from 5 unit-squares in the tangential direction; the "gradient" is thus $m = \frac{1}{5}$. In general, a logarithmic spiral with gradient m is transformed, under a circle-reflection in a reference-circle concentric with the spiral center, into a logarithmic spiral with a gradient $(-m)$. In a logarithmic spiral the angle to the radial gridlines is always the same, namely $\beta = \text{arc cot } m$.

A logarithmic spiral thus follows a constant course β with respect to the grid-center. Such curves appear in the real world. For a few insects, for example, for bees, the lateral placing of the immovable eyes has the consequence that the insects fly towards an object in view, for example a flower, with a systematic angular error β; the flight curve is thus a logarithmic spiral. Also, in aviation and navigation, constant courses are adopted; the curves thus arising, the so-called **loxodromes,** are, in view of the sphericity of the earth, not logarithmic spirals, but they can be approximated by such curves near the poles.

Figure 27a shows a loxodrome, in which, for two unit-squares in the easterly direction, there occurs a unit-square in the northerly direction. It cuts each meridian at an angle $\arctan 2 \approx 63.43°$, and so follows a constant course of $63.43°$. Figure 27b shows the same loxodrome as viewed from the north. In Figure 28 (from [32, p. 318]) we recognize the same loxodrome as the basis of the design.

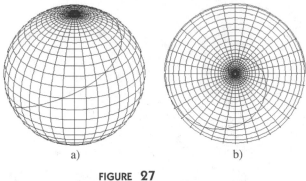

a)   b)

FIGURE 27
Loxodrome with constant course

Question 2.3 How long is a logarithmic spiral from reference-circle to its innermost point?

Question 2.4 Which curves are given by the polar equations $r(\phi) = a\phi + b$, $r(\phi) = \frac{a}{\phi}$?

FIGURE 28
M. C. Escher: *Spherical surface with fish*, July 1958. © 1997 Cordon Art–Baarn–Holland. All rights reserved.

Question 2.5 How long is the loxodrome of Figure 27 from South Pole to North Pole?

2.5 IMAGE OF A STRAIGHT LINE

We study the image of a straight line g under a circle-reflection in a reference-circle h. To this end we choose a Cartesian coordinate system such that h is given by the equation $x^2 + y^2 = r_0^2$ and g is given by the equation $x = c$ (Figure 29a).

The line g is then given in polar coordinates by the equation $c = r \cos \phi$. Since $rr' = r_0^2$, the equation for the image figure, in polar coordinates, is

$$r' = \frac{r_0^2}{c} \cos \phi.$$

The points which satisfy this equation lie on the circle having the line-segment joining $M(0,0)$ and $B'(\frac{r_0^2}{c}, 0)$ as diameter (see Figure 29a). The image g' of g is thus a circle through M itself. The center M of h can, however, be interpreted as the image of the ***point at infinity*** on g. The tangent to g' at M is parallel to g.

Remark The image of a straight line g is, of course, also a circle if G intersects the reference-circle (Figure 29b). In this case g' may simply be described as the circle through M and the two fixed points F_1 and F_2.

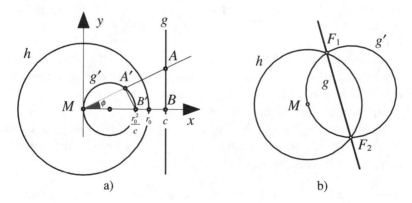

FIGURE 29
Image of a straight line

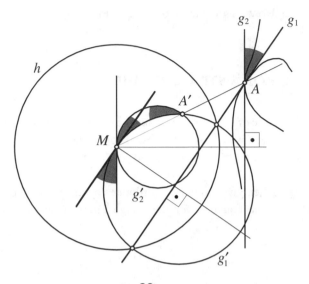

FIGURE 30
The angle remains fixed

Question 2.6 Are there exceptional straight lines whose image under a circle-reflection is not a circle?

We can now show that the angle is preserved under a circle-reflection in h. The angle of intersection of two curves is defined as the angle of intersection of their tangents at the point of intersection A (Figure 30). But this angle is equal to the angle between the image circles of the two tangents to the curves at M and is thus equal to the angle at the image point A'.

A mapping under which angles are preserved is called **angle-preserving** or **conformal**. Conformal mappings play an important role in cartography in the construction of angle-preserving maps.

Circle-reflections are conformal mappings.

In particular, right angles are preserved under conformal mappings. The image of a Cartesian square lattice is thus itself an orthogonal network, consisting of orthogonal circles through M (Figure 31).

Question 2.7 How did the gothic quatrefoil window in the center of Figure 31 arise?

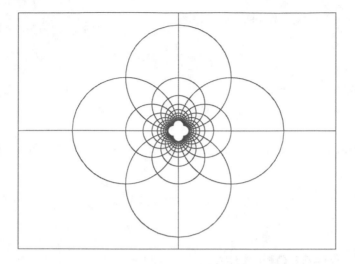

FIGURE **31**
Image of a square lattice

2.6 REPRESENTATION IN CARTESIAN COORDINATES

We consider a Cartesian coordinate system with the center M of the reference-circle as origin (Figure 32). With the notations $A(x, y)$ and $A'(x', y')$, we can write $r = \sqrt{x^2 + y^2}$, $r' = \sqrt{x'^2 + y'^2}$ and $\frac{x'}{x} = \frac{r'}{r}$.

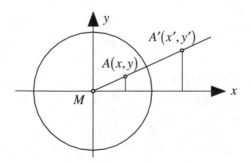

FIGURE **32**
Representation in Cartesian coordinates

Since $rr' = r_0^2$, it further follows that

$$\frac{x'}{x} = \frac{r'}{r} = \frac{r_0^2}{r^2} = \frac{r_0^2}{x^2 + y^2}.$$

This yields the mapping equations

$$x' = r_0^2 \, \frac{x}{x^2 + y^2}, \quad y' = r_0^2 \, \frac{y}{x^2 + y^2}.$$

Remark For the inverse mapping we have the symmetric mapping equations

$$x = r_0^2 \, \frac{x'}{x'^2 + y'^2}, \quad y = r_0^2 \, \frac{y'}{x'^2 + y'^2}.$$

2.7 IMAGE OF A CIRCLE

The image of a circle k concentric with the reference-circle h is again a circle k' concentric with h; for the radii ρ, ρ' of the circles k, k' we have $\rho\rho' = r_0^2$. A circle k through the center M of h has as image a straight line. We study now the image of a circle k which is neither concentric with h nor passes through M (Figure 33). In an appropriate Cartesian coordinate system such a circle k with center at $(u, 0)$ and radius ρ is given by the equation $(x - u)^2 + y^2 = \rho^2$. By substituting from the mapping equations for the

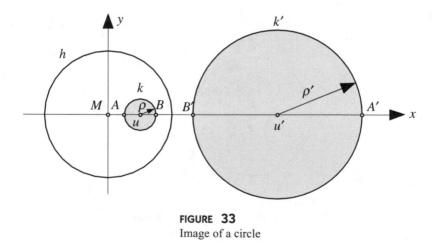

FIGURE 33
Image of a circle

inverse mapping, we obtain from this for the image figure k' the equation

$$\left(r_0^2 \frac{x'}{x'^2 + y'^2} - u \right)^2 + \left(r_0^2 \frac{y'}{x'^2 + y'^2} \right)^2 = \rho^2.$$

We manipulate this into the form

$$r_0^4 - 2ur_0^2 x' + u^2(x'^2 + y'^2) = \rho^2(x'^2 + y'^2).$$

This is again the equation of a circle; it can be brought into the form

$$\left(x' - \frac{ur_0^2}{u^2 - \rho^2} \right)^2 + y'^2 = \left(\frac{r_0^2 \rho}{u^2 - \rho^2} \right)^2$$

The image k' of the circle k is thus a circle with center $(u', 0)$ and radius ρ', where

$$u' = \frac{ur_0^2}{u^2 - \rho^2}, \quad \rho' = \frac{r_0^2 \rho}{u^2 - \rho^2}.$$

Remark We have $uu' = \frac{u^2 r_0^2}{u^2 - \rho^2} \neq r_0^2$; this means that the center of one circle is *not* mapped onto the center of the other.

Under a straightline reflection, a circle which cuts the axis of reflection orthogonally is fixed (Figure 34a).

The same is true with a circle-reflection; a circle k which cuts the reference-circle h orthogonally is a fixed circle (Figure 34b). In this situation we have, from Pythagoras' Theorem, $u^2 - \rho^2 = r_0^2$. Hence

$$u' = \frac{ur_0^2}{u^2 - \rho^2} = u; \quad \rho' = \frac{r_0^2 \rho}{u^2 - \rho^2} = \rho.$$

Remark Thus, with circle-reflections, circles and straight lines are transformed, according to the situation, into circles and/or straight lines. It is

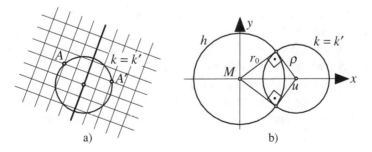

FIGURE 34
Fixed circles

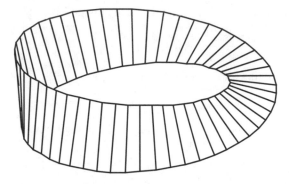

FIGURE 35
The Möbius band

therefore meaningful to subsume both ideas under one all-embracing idea. The concept of **Möbius circle** brings together circles and straight lines; a Möbius circle is one or the other. Under a circle-reflection the image of a Möbius circle is always a Möbius circle. The name Möbius circle reminds us of A. F. Möbius.

August Ferdinand Möbius (born 17/11/1790 in Schulpforta, died 26/9/1868 in Leipzig) was, after studies at Leipzig University and study leave in Göttingen and Halle, called in 1816 on the recommendation of Gauss to Leipzig. In 1820 he became Director of the Observatory, and in 1844 full professor for astronomy and mechanics. He contributed decisively to the education of Gymnasium (academic high school) teachers in the kingdom of Saxony and was heavily involved in the new orientation of geometry in the first half of the 19th century. He is also the discoverer of the Möbius band (Figure 35). A Möbius band is easily constructed from a strip of paper, into which one introduces a single twist before sticking the ends together. The Möbius band has some interesting topological properties. It is a "one-sided surface", that is, it has no "inside" and "outside". If we start anywhere on the apparent outside to color the band we find ourselves, as we continue the work, suddenly on the apparent inside. The ants of Figure 36 (from [32, p. 324]) pass without problem from the apparent outside to the apparent inside, without having to climb over the edge of the strip.

Question 2.8 What happens if we try to cut a Möbius band in two along the center line?

FIGURE 36
M. C. Escher: Möbius strips II (Red
Forest Ants), 1963. © 1997 Cordon
Art–Baarn–Holland.

The concept of circle-reflection can be extended to sphere-reflection in
3-dimensional space. Then the inside of the sphere is mapped onto the outside
and conversely.

Question 2.9 How does a mathematician catch a lion?

2.8 SQUARE-REFLECTION

The idea of reflecting from the inside to the outside and vice versa can be
carried over to other closed figures. For the inspiration of this idea I must
thank Georg Schierscher. In Figure 37 a reflection in a square of side length
2 is represented. The "reflection in a square" is defined as follows: the point
A' lies on the same ray emanating from the center M of the square as the

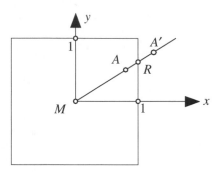

FIGURE **37**
Reflection in a square

original point A. The point R, with now a variable polar distance, which we still denote by r_0, is the intersection of the ray with the square. For the polar distances r and r' of A and A', respectively, we have the relationship $rr' = r_0^2$.

Question 2.10 Express the mapping equations of this reflection in a square in Cartesian coordinates. How does the image of the square grid parallel to the axes look? What is the image of the square grid turned through 45°?

2.9 OTHER REFLECTIONS

Question 2.11 What problem presents itself if one reflects in a parabola p of the parabolic network of Figure 38 by "counting the unit squares"?
Further "near-reflections" are considered in [4].

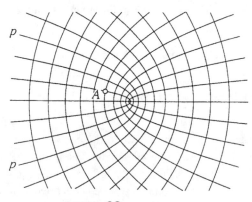

FIGURE **38**
Reflection in a parabola?

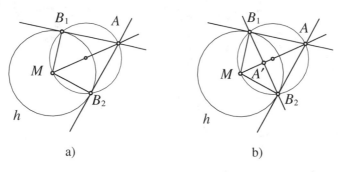

FIGURE **39**

The point A lies outside h

Answers to Questions

Answer 2.1 The construction proceeds in the reverse sequence according to Figure 39.

Answer 2.2 Precisely the points of the reference-circle h are the fixed points.

Answer 2.3 A path-length ds on a logarithmic spiral has, in view of the constant course, a radial variation given by $dr = ds \cos \beta$. Thus the length of the logarithmic spiral from the reference-circle h with radius r_0 to its innermost point is given by $s_0 = \frac{r_0}{\cos \beta}$.

Answer 2.4 A linear function $r(\phi) = a\phi + b$ in polar coordinates yields an ***archimedean spiral*** (Figure 40a). For an archimedean spiral the radial difference between two revolutions is always the same, namely, $r(\phi+2\pi)-r(\phi) = 2a\pi$. Such spirals arise from the rolling up of material of constant thickness (rolls of carpet, rolls of film, rolls of toilet paper). The mirror-image of an archimedean spiral in a circle-reflection is no longer an archimedean spiral.

The function $r(\phi) = \frac{a}{\phi}$ yields a ***hyperbolic spiral,*** which, in view of its appearance, is also known as a ***bishop's staff*** or ***crosier*** (Figure 40b). We see such a hyperbolic spiral when we look at a screw in the direction of its axis, for example, by looking at the central shaft of a winding staircase.

Answer 2.5 According to the Theorem of Pythagoras, the arc length for a unit length in the northerly direction is $\sqrt{5}$ times this unit length (Figure 41). The entire loxodrome length is thus $\sqrt{5}$ times the meridian length. For the loxodrome with constant course β the length, from Pole to Pole, is $\frac{1}{\cos \beta}$ times the meridian length. Thus it is also clear that, in general, the loxodrome does not constitute the shortest path on the spherical

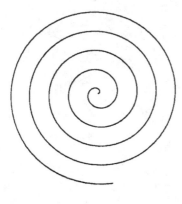

a)   b)

FIGURE 40
Archimedean and hyperbolic spirals

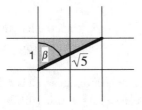

FIGURE 41
Arc length of the loxodrome

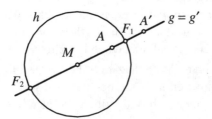

FIGURE 42
Image of a straight line through the center

surface. The shortest path is given by arcs of great circles. Nevertheless on sea journeys the loxodrome is preferred on practical grounds (constant course) when the difference in length is not too great.

Answer 2.6 A straight line g through the center M is a fixed line, though not a line of fixed points. It has only the two diametrical fixed points F_1 and F_2. As to the other points, the inner and outer points are exchanged; the point M is mapped to the point at infinity, and conversely (Figure 42). The image straight line g' can be viewed as a circle of infinite radius.

Answer 2.7 Figure 31 is the image of a bounded Cartesian square network; the region runs in both dimensions from -12 to $+12$. The gothic quatrefoil window is the image of the boundary square with $x = \pm 12$ and $y = \pm 12$ under reflection in a circle.

Answer 2.8 The Möbius band cannot literally be cut in two along its center line. If attempted, there results a *single* band, twice as long and with a double twist. Try it!

Answer 2.9 He (or she) sits in the center of an empty lion cage and comes out through a reflection in a sphere. Then he (or she) is outside and the lion is inside. Indeed, all lions are then in the cage. Moreover, the mathematician, and the rest of humanity, no longer have their hearts in the right place. (Translator's note: the lion cage is assumed spherical.)

Answer 2.10 For the mapping equations we need a separation of cases. If $|y| \leq |x|$ we have

$$x' = \frac{x}{x^2}; \quad y' = \frac{y}{x^2},$$

and if $|y| > |x|$ we have

$$x' = \frac{x}{y^2}; \quad y' = \frac{y}{y^2}.$$

Figure 43a shows the image of the gridline $y = $ const. Instead of the circles of

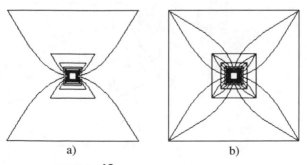

FIGURE 43

Image of the square network parallel to the axes

FIGURE 44
Image of the square network turned through 45°

Figure 31 we obtain curves which are put together from quadratic parabolae and line segments. Figure 43b shows the image of the entire square grid parallel to the axes. Figure 44 shows the image of the square grid turned through 45°.

Answer 2.11 The mirror image is not uniquely defined; there is an "upper" and a "lower" solution (Figure 45).

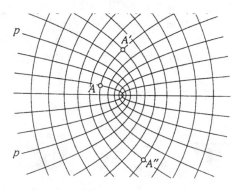

FIGURE 45
The mirror image is not uniquely defined.

CHAPTER **3**

Symmetric Procedure

In this section symmetry is employed, in exemplary fashion, as a ***method.***
It is shown that reasoning, construction and calculation are often simplified
if the approach, as in building a tunnel, is made "from both ends inwards".
The subject matter treated in this section includes central points, in the sense
of centers of gravity (or centroid) of triangles and quadrilaterals.

3.1   CENTER OF GRAVITY IN THE TRIANGLE

The center of gravity S of a triangle is the common intersection point of
the 3 medians (Figure 46). This center of gravity is both "vertex center
of gravity" (Figure 46a) and "surface (or area) center of gravity" (Figure
46b) of the triangle. "Vertex center of gravity" means that S is the center of
gravity of 3 equal point-masses at the 3 vertices (Figure 46a). If we denote

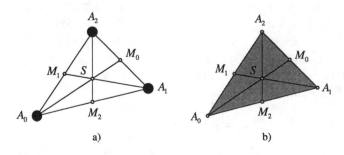

a)   b)

FIGURE 46
The center of gravity of a triangle

31

the position vector of the vertex A_i by $\vec{a}_i$ and the position vector of the center of gravity S with $\vec{s}$, then this means that

$$\sum_{i=0}^{2}(\vec{a}_i - \vec{s}) = \vec{0},$$

from which the formula

$$\vec{s} = \frac{1}{3}(\vec{a}_0 + \vec{a}_1 + \vec{a}_2)$$

follows.

"Surface center of gravity" means that S is the center of gravity of a homogeneous triangular plate (Figure 46b). For computing purposes this reads

$$\iint\limits_{\text{triangle}} (\vec{x} - \vec{s})\, dx\, dy = \vec{0},$$

which leads us to the same formula for $\vec{s}$.

Question 3.1 What is the "edge center of gravity" of a triangle?

Question 3.2 Each median bisects the triangular surface. Do the other straight lines through the center of gravity bisect the triangular surface?

3.2 CENTERS OF GRAVITY IN THE QUADRILATERAL

In the quadrilateral the vertex center of gravity and the surface center of gravity no longer coincide (see [26], [43]). For the construction of the vertex center of gravity, we may think of the vertices as divided into two "dumbbells", whose centers of gravity are the midpoints of the linking segments. The vertex center of gravity can thus be constructed as the midpoint of the segment $\overline{M_1 M_3}$ (Figure 47), or equally well as the midpoint of the segment $\overline{M_0 M_2}$ (Figure 48).

For the position vector $\vec{e}$ of the vertex center of gravity E we have, by analogy with the center of gravity of a triangle:

$$\vec{e} = \frac{1}{4}(\vec{a}_0 + \vec{a}_1 + \vec{a}_2 + \vec{a}_3).$$

To determine the surface center of gravity F of the quadrilateral, we divide it by a diagonal into two triangles and construct for each triangle its center of gravity (Figure 49a).

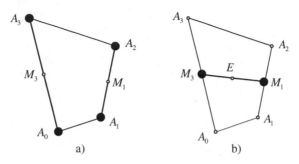

FIGURE 47
Vertex center of gravity in the quadrilateral

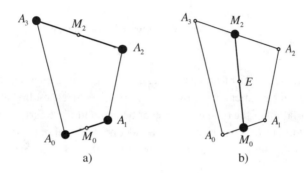

FIGURE 48
Symmetric procedure

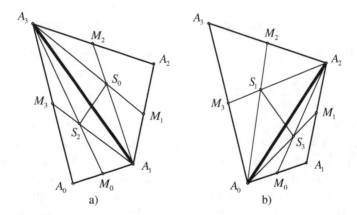

FIGURE 49
Subdivision into triangles

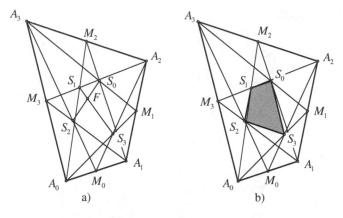

FIGURE 50
Surface center of gravity of a quadrilateral

The surface center of gravity F must now lie on the segment $\overline{S_0 S_2}$ and must divide the segment in the ratio inverse to the corresponding area ratios. This calculation of the ratio can however be avoided if we carry our the analogous "symmetric" construction with a second diagonal of the quadrilateral (Figure 49b). The surface center of gravity is then the intersection of the segments $\overline{S_0 S_2}$ and $\overline{S_1 S_3}$ (Figure 50a).

Remark On the basis of Figure 50a we conjecture that the quadrilateral $S_0 S_1 S_2 S_3$ with vertices the 4 triangle centers of gravity is similar to the original quadrilateral $A_0 A_1 A_2 A_3$ (Figure 50b). Notice that, in both Figure 50a and Figure 50b, the diagonals have been omitted for clarity.

This conjecture is indeed valid. From $\vec{s}_0 = \frac{1}{3}(\vec{a}_1 + \vec{a}_2 + \vec{a}_3)$ it follows that

$$\frac{1}{4}\vec{a}_0 + \frac{3}{4}\vec{s}_0 = \frac{1}{4}(\vec{a}_0 + \vec{a}_1 + \vec{a}_2 + \vec{a}_3) = \vec{e}.$$

This means that the vertex center of gravity E lies on the segment $\overline{A_0 S_0}$ and divides it in the ratio $3:1$. The same holds for the remaining segments $\overline{A_i S_i}$. Thus the central dilatation with center E and stretch factor $-\frac{1}{3}$ maps the starting quadrilateral $A_0 A_1 A_2 A_3$ onto the quadrilateral $S_0 S_1 S_2 S_3$. The surface center of gravity F is the intersection of the diagonals of the quadrilateral $S_0 S_1 S_2 S_3$ and thus the image, under this dilatation, of the intersection D of the diagonals of the quadrilateral $A_0 A_1 A_2 A_3$. Thus we have: In an arbitrary quadrilateral $A_0 A_1 A_2 A_3$ the intersection of the

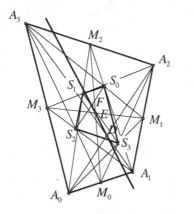

FIGURE 51
Collinearity

diagonals D, the vertex center of gravity E and the surface center of gravity F are collinear, and E divides $\overline{DF}$ in the ratio $3:1$ (Figure 51).

Question 3.3   What about the edge center of gravity of a quadrilateral?

Answers to the Questions

Answer 3.1   We are looking for the center of gravity of a triangular model that consists of edges of homogeneous cross-section, for example, made of equally thick wire. To determine this edge center of gravity, we look first for the center of gravity of the two edges $a_0 = \overline{A_1 A_2}$ and $a_1 = \overline{A_2 A_0}$ (Figure 52a).

Each of these edges a_i has its midpoint M_i as center of gravity, so the center of gravity G_2 of the two edges lies on the segment $\overline{M_0 M_1}$ and, in fact, G_2 divides this segment in inverse proportion to the masses, and thus

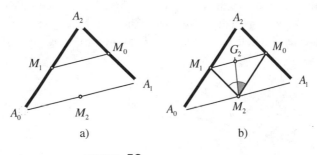

a)   b)

FIGURE 52
Center of gravity of two edges

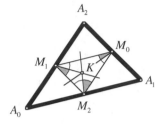

FIGURE **53**
Edge center of gravity of a triangle

the lengths, of the sides in question. Thus G_2 is the intersection of $\overline{M_0M_1}$ and the bisector of the angle $M_0M_2M_1$, since the segments $\overline{M_0M_2}$ and $\overline{M_1M_2}$ have the same length-ratios as a_1 and a_0, and the angle bisector in the triangle $M_0M_1M_2$ divides the opposite side in this ratio. Since M_2 is the center of gravity of the third edge, the sought-for edge center of gravity must lie on the segment $\overline{G_2M_2}$. We could now work with the mass-distribution whereby we put the union of the masses of a_0 and a_1 at G_2 and the mass a_2 at M_2. But it is simpler to reason by symmetry. The looked-for edge center of gravity K lies not only on the segment $\overline{G_2M_2}$, but also on the similarly constructed segments $\overline{G_0M_0}$ and $\overline{G_1M_1}$, and is therefore the common point of intersection of these 3 segments. The edge center of gravity K is thus the intersection of the bisectors of the angles or the incenter of the triangle $M_0M_2M_1$ whose vertices are the midpoints of the edges (Figure 53). The point K is sometimes called the Spieker center of the triangle [24a].

Answer 3.2 No. The simplest counterexample is a line through the center of gravity parallel to a side (Figure 54). From the given grid it is plain that the trapezoidal region contains 5 grid units, while the triangular region contains only 4.

Answer 3.3 To determine the edge center of gravity K of the quadrilateral $A_0A_1A_2A_3$, we first seek the center of gravity K_0 of the two edges of the quadrilateral incident with A_0 (Figure 55a). By analogy with the procedure in the case of the edge center of gravity of the triangle, we locate the point K_0 at the intersection of the diagonal $\overline{M_0M_3}$ of the parallelogram $A_0M_0P_0M_3$

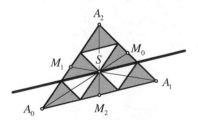

FIGURE **54**
No area-bisection

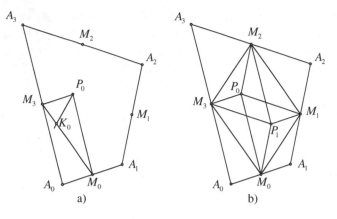

FIGURE **55**
Center of gravity of two edges

with the bisector of the angle at P_0. In the same way we construct the center of gravity K_2 of the two other edges of the quadrilateral.

Remark We determine thereby that the point P_2 coincides with P_0, as may also be proved as follows: We denote by $\vec{p}_0$ the position vector of the point P_0. Then

$$\vec{p}_0 = \vec{a}_0 + \frac{1}{2}(\vec{a}_1 - \vec{a}_0) + \frac{1}{2}(\vec{a}_3 - \vec{a}_0) = \frac{1}{2}(\vec{a}_1 + \vec{a}_3).$$

Thus P_0 is the midpoint of the line-segment $\overline{A_1 A_3}$. Further, we have

$$\vec{p}_2 = \vec{a}_2 + \frac{1}{2}(\vec{a}_2 - \vec{a}_2) + \frac{1}{2}(\vec{a}_3 - \vec{a}_2) = \frac{1}{2}(\vec{a}_1 + \vec{a}_3)$$

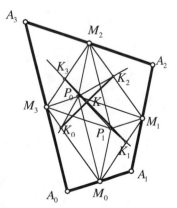

FIGURE **56**
Edge center of gravity of a quadrilateral

and P_2 is just as well the midpoint of the line-segment $\overline{A_1 A_3}$ and coincides with P_0. In the same way, P_1 and P_3 coincide, and the six points $M_0 M_1 M_2 M_3 P_0 P_1$ appear as the vertices of an affine projection of a regular octahedron (Figure 55b). This is a generalization of the fact that the 4 midpoints $M_0 M_1 M_2 M_3$ form a parallelogram.

The edge center of gravity we seek must lie on the line-segment $\overline{K_0 K_2}$ and also, by symmetry, on the line-segment $\overline{K_1 K_3}$; it is therefore the point of intersection of these two segments (Figure 56). The edge center of gravity K does not lie on the line passing through the vertex center of gravity E and the surface center of gravity F.

CHAPTER 4

Parquet Floors, Lattices and Pythagoras

4.1 PARQUET FLOORS[2]

Figure 57 shows the 3 simplest patterns which consist of regular polygons.

The patterns of squares or equilateral triangles can be displayed in chessboard fashion with two colors; with the hexagonal pattern three colors are necessary if no two neighboring hexagons are to have the same color.

These parquets contain very many symmetries, provided in studying their symmetries the parquets are regarded as "infinitely large." On a chessboard, then, all axes of symmetry of an individual square are also axes of symmetry of the entire pattern. If the black-and-white coloring is ignored then the lines containing edges of each individual square are axes of symmetry of the

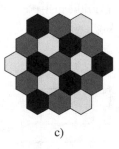

a)   b)   c)

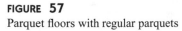

FIGURE 57
Parquet floors with regular parquets

---

[2] These are referred to in many specialized texts as "tilings" (see, e.g., [18]).

whole pattern as well. Further, the centers of the squares are centers of 4-fold symmetry of the parquet, that is, the parquet can be turned through any multiple of 90° about such a point—i.e., a multiple of a quarter of a full rotation—without anything changing in the appearance of the pattern. If attention is paid to the black-and-white coloring, the vertices of the square are centers of 2-fold symmetry, but if the coloring is ignored, they are centers of 4-fold symmetry. Finally, if the color is ignored, then the midpoints of the sides of a square are centers of 2-fold symmetry.

The most important symmetry—above all in reference to the technical applications and manufacture of parquets—is, however, translation symmetry. The chessboard can be slid two square-lengths (or a multiple of this), taking account of the coloring, either horizontally or vertically, without its appearance changing.

The edges of the parquets of Figure 58 form a *lattice* with *lattice lines;* if we confine attention to the vertices we speak of *lattice points* (Figure 58). The lattice points of a square lattice are exactly the points that can be given integer coordinates in a Cartesian coordinate system.

Question 4.1 Are there also parquets with other regular polygons?

Figure 59 shows a variation of the chessboard pattern; the tiles of the parquet arise from a deformation of a square, wherein the alteration of one side of the square is compensated for on the opposite side (Figure 59b). This parquet, however, contains far fewer symmetries than the chessboard pattern; only the translation symmetries survive, the reflection and rotation symmetries disappear.

Such patterns can easily be drawn with graphic software, which recognizes a "grid-catcher". It is enough to draw a single tile of the parquet in such a way that the vertices of the fundamental square fit into the grid. This tile can

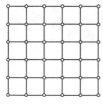   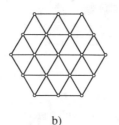   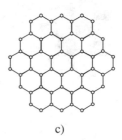

a)   b)   c)

FIGURE 58
Lattice lines and lattice points

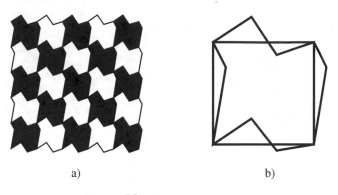

a)   b)

FIGURE 59
Variation of the chessboard pattern

then be copied arbitrarily often and the copies can accordingly be fitted into the grid.

With square pieces, and with equilateral triangles, parquets with displaced tiles can also be laid down (Figure 60); with square parquets we then need three different colors. Such displacement leads to a loss of possible symmetries. With a parquet of regular hexagons a displacement is not possible.

Parquets with designs on the interior of individual parquet tiles have been used by the well-known Dutch graphic artist Maurits Cornelis Escher (1898–1972) in several variations (cf. [12], [13], [32] and [41]).

Figure 61 shows parquets with general triangles and quadrilaterals as parquet tiles. The triangular parquet is here simply an affine distortion of the parquet with equilateral triangles (Figure 57b); the quadrilateral parquet, on the other hand, cannot be regarded as an affine distortion of

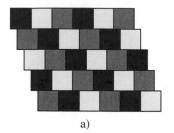 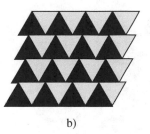

a)   b)

FIGURE 60
Parquets with "displacement"

a)   b)

FIGURE **61**
General triangles and quadrilaterals

the chessboard pattern. An affine distortion of the chessboard pattern must consist of parallelograms.

Question 4.2 There are many different possibilities for laying out a parquet without displacement, using rhombi with an acute angle of 60°. Describe two of these.

Question 4.3 Can a parquet with a general hexagon be laid out?

Interesting parquets can be obtained when tiles of different parquet shapes are combined. Figure 62 shows an example with equilateral triangles and squares.

Question 4.4 Are there further combinations of equilateral triangles, squares, and regular hexagons?

Occasionally, with apparently irregular parquets, there turn up new unexpected regularities and symmetries. In Figure 63 there is a general triangle combined with three equilateral triangles of different sizes. We may

FIGURE **62**
A combination of equilateral triangles and squares

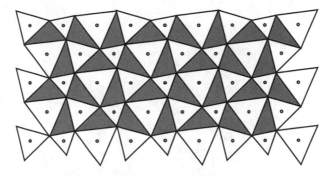

FIGURE **63**
Various triangles

determine that the midpoints of the equilateral triangles, for their part, lie on a regular triangular lattice. See [38a, 38b, 38c] for more on this topic.

Question 4.5 Why is this so?

Question 4.6 What could the combination of a general parallelogram with squares produce?

Specific literature on parquets—in particular on questions of their enumeration and classification and on the group-theoretical interpretation of the different possibilities for parquets—may be found in [5], [6], [14], [18] and [31]. In [7] the aesthetic aspect is emphasized.

4.2 PARQUETS AND PYTHAGORAS

In Figure 64, 3 different parquets are recognizable. These 3 parquets "fill" the two cathetus-squares[3] CBA_1A_2 and ACB_1B_2 as well as the hypotenuse-square BAC_1C_2 of the right-angled triangle ABC. The word "fill" is here to be understood as meaning that an arbitrary deficiency in one side of a cathetus-square is balanced by an equal surfeit in the opposite side of the square. The parquet and the corresponding square thus have equal area. The parquet in the cathetus-square ACB_1B_2 consists of 48 congruent tiles, the parquet in the cathetus-square CBA_1A_2 is made up of 24 squares

---

[3] A cathetus is one of the shorter sides of a right-angled triangle. Most English texts use the word "leg."

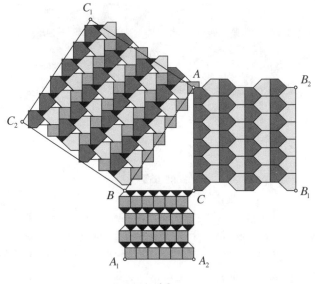

FIGURE 64
Forming parquets

and 48 right-angled isosceles triangles. The parquet in the hypotenuse-square BAC_1C_2 contains precisely the tiles of the two cathetus-squares in the appropriate size, shape and number. Thus we have found a "proof by parquets" of Pythagoras' Theorem.

Is this proof by parquets valid only for the right-angled triangle of Figure 64 with the short sides in the ratio $a:b = 2:3$, or is it valid for any right-angled triangle? And are there also other choices of parquets yielding a proof of Pythagoras' Theorem?

The two cathetus-parquets are related to each other. This can be seen as follows: In the parquet for cathetus a (Figure 65a) we mark all vertices

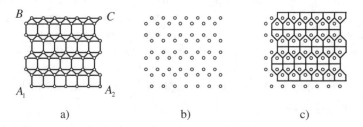

FIGURE 65
Connection between the two parquets

(Figure 65b) and then draw, for neighboring vertices, the right-bisector (Figure 65c). There arises thereby a parquet similar to the second cathetus-parquet; but it has a different size and is turned through a right angle.

Remark The parquet of Figure 65c is called the ***Dirichlet parquet*** of the point-set of Figure 65b. (cf [6, p. 354 ff].

Question 4.7 How should the children be assigned to schools in a city with several schools, so that each child has a shortest possible route to school?

Question 4.8 If we draw the Dirichlet parquet of the second cathetus-parquet, we do not come back to the first cathetus-parquet. Are there parquets which are equivalent to the Dirichlet-parquet of their Dirichlet-parquet? Are there parquets which are equivalent to their Dirichlet-parquets?

4.3   CONSTRUCTION OF A PROOF-DIAGRAM

We will see that, for an arbitrary right-angled triangle, there are infinitely many "parquet-proofs". We begin with a parquet with convex parquet pieces with straight-line boundaries, which have the following property: In each parquet piece a "principal point" may be picked out, such that the segment joining the principal points of neighboring pieces is orthogonal to their common boundary; conversely, just one such linking segment runs over each such stretch of the common boundary. Figure 66 shows an example.

Remark Such a parquet can contain irregularities. The parquet need not be a Dirichlet-parquet; moreover, the second parquet defined by the principal points (the so-called ***dual*** parquet) need not be a Dirichlet-parquet either. Conversely, however, every Dirichlet-parquet fulfills the condition

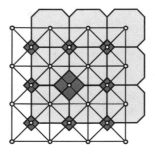

FIGURE 66
Example

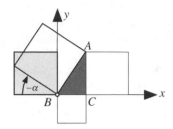

FIGURE 67
Construction of the stretch-rotation

formulated above, the principal points being precisely the points on which the construction of the Dirichlet-parquet is based.

This parquet, with its principal points, we use as a cathetus-parquet for the cathetus b of the right-angled triangle. To find the appropriate transition to the hypotenuse-parquet, we need a stretch-rotation with angle of rotation $-\alpha$ and stretch-factor $\frac{c}{b} = \frac{1}{\cos \alpha}$. Figure 67 shows how the properly shifted cathetus-square of the cathetus b is carried over into the square on the hypotenuse. Analytically, this stretch-rotation is described by the mapping equations

$$\overline{x} = x + \frac{a}{b} y$$

$$\overline{y} = -\frac{a}{b} x + y.$$

We now map the principal point P_k of the parquet—but without the parquet tile piece to which it belongs—with this stretch-rotation to the point $\overline{P}_k$. The parquet tiles themselves we move by translations in such a way that their principal points come to lie on the corresponding points $\overline{P}_k$. Figure 68 shows on the left a section from the cathetus-parquet with five tiles and, on the right, the situation after the process described above has been carried out. Between the tiles there is now a gap which is similar to the tile of the parquet defined by the principal points.

Question 4.9 How do we get from the "gap" to the similar pentagon $\overline{P}_0\overline{P}_1\overline{P}_2\overline{P}_3\overline{P}_4$?

We can fill these holes with suitable parquet tiles which, tile by tile, are similar to the parquets defined by the principal points. Thus we obtain the second cathetus-parquet.

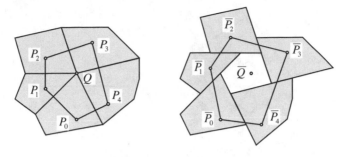

FIGURE 68
Movement of the parquet pieces

Question 4.10 What does the parquet with squares look like?

In the example of Figure 69a the cathetus-parquet of the cathetus b is built from octagons and squares. Figure 69b shows a cathetus-parquet with a "disturbance", which, however, with our procedure, is absorbed without disturbance of the other cathetus-parquet.

With a parquet tiling with non-convex tiles it is not possible to determine principal points. Nevertheless it is possible in a few cases to build a proof-figure for Pythagoras' Theorem. Figure 70 shows an example. But, in general, the non-convex tiles will overlap each other in the attempt to accommodate them in the hypotenuse-square.

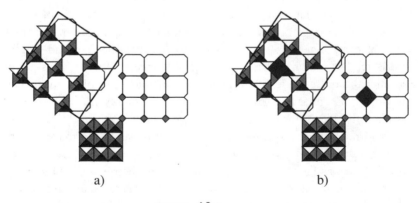

a)   b)

FIGURE 69
Various parquet pieces

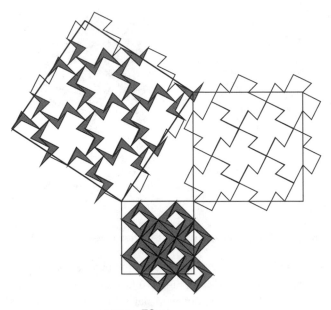

FIGURE 70
Non-convex parquet pieces

4.4   OTHER CATHETUS-FIGURES

In our considerations thus far we have tacitly assumed that the cathetus-squares can, without difficulty, be "filled" by the parquet. This is not always possible; for example, a square cannot be parqueted with equilateral triangles so that everything works with respect to the area-measure and length-measure. The reason is that in an equilateral triangle the height is not in a rational ratio to the length of a side. An analogous problem arises with a parquet of regular hexagons. The problem can, however, be solved if we work with other cathetus- and hypotenuse-figures. Pythagoras' Theorem is valid not only for squares—though it is most simply formulated with squares—but for any triple of figures, mutually similar, whose length-ratios are the same as the ratios $a : b : c$ of the sides of the right-angled triangle. The three figures have then the area-ratios $a^2 : b^2 : c^2$, that is, the area-sum of the cathetus-figures is the area of the hypotenuse-figure. In the 3-kings[4] figure

---

[4]   The reader might prefer the word "crown" to "kingdom."

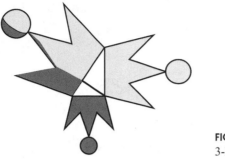

FIGURE 71
3-kings figure

of Figure 71, the united kingdom on the hypotenuse is equal in area to the sum of the cathetus-kingdoms.

Question 4.11 Figure 72 shows a right-angled triangle with equilateral triangles erected on the hypotenuse and each cathetus. In this situation, how can the hypotenuse-triangle be subdivided in such a way as to illustrate the equality of its area with that of the union of the two cathetus-triangles? How can Pythagoras' Theorem be demonstrated by parqueting?

Figures 73 and 74 use regular hexagons on the sides of the triangle. Thus are parquetings from the "family of triangles" possible.

Another approach to "Pythagorean parquets" is to be found in [16, p. 254 ff].

4.5 COVERING OF LATTICE POINTS

Let us place two transparencies, consisting of congruent square lattices with inscribed lattice points, arbitrarily one on top of the other (Figure 75); there arises thereby in the domain of overlap a new pattern which for its part seems to exhibit certain symmetries. This effect we will designate as a ***Moiré-effect;***

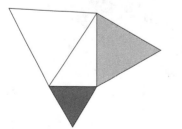

FIGURE 72
Equilateral triangles

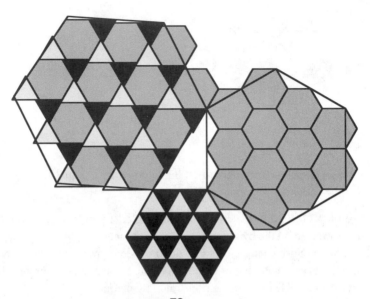

FIGURE 73
Triangles and hexagons

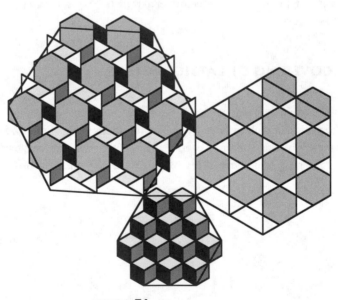

FIGURE 74
Triangles, hexagons and rhombi

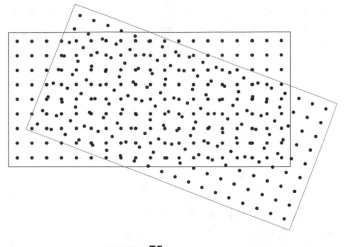

FIGURE 75
Overlapping square lattices

such an effect arises in general as an optical effect from the superimposition of two congruent patterns, for example, by looking through two perforated metal sheets. One can generate the effect oneself by laying two transparencies with the same pattern one on top of the other—but then the two grids must be exactly alike—or by overlapping two copies of a virtual transparency in some graphic software.

Question 4.12 What arises from a parallel superimposition of two congruent square lattices?

We study now the Moiré-effect in the superimposition of a square lattice and its mirror image. If we choose as axis of reflection an axis of symmetry of the square lattice, then obviously no effect is visible. A square lattice, regarded as infinitely large, contains three types of axis of symmetry (Figure 76): a) Sides of the lattice squares, b) diagonals of the lattice squares and c) center-lines of the lattice squares.

In what follows we denote the set of lattice points of the square lattice by G. In lattice geometry the notion of ***primitive segment*** is often used: A primitive segment $\overline{AB}$ in a square lattice G has 2 lattice points A and B as endpoints, but otherwise contains no lattice points. Each side or diagonal of a square lattice is primitive. All other primitive segments in the square lattice G have a ***support triangle*** with coprime cathetus-lengths u and v (Figure 77).

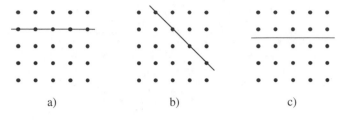

a)   b)   c)

FIGURE 76

The three types of axis of symmetry

By means of a primitive segment with a support triangle a new square lattice J with $J \subset G$ can be defined (Figure 78).

In a square lattice G let $\overline{AB}$ be a primitive segment with support-triangle cathetus u and v (Figure 77) and s the straight line AB. Further, let $G' := \sigma_s(G)$ be the image of G under reflection in s, and $H := G \cap G'$ the set of points common to G and G'; H is thus the subset of G symmetrical with respect to the axis of symmetry s.

Then the following theorem holds [49].

H is a square lattice with edge-length e, where

$$e = \sqrt{u^2 + v^2} \quad \text{if} \quad u \not\equiv v \bmod 2,$$

$$e = \frac{1}{\sqrt{2}} \sqrt{u^2 + v^2} \quad \text{if} \quad u \equiv v \bmod 2.$$

In the case $u \not\equiv v \bmod 2$, the axis of reflection s is a lattice-line in H; in the case $u \equiv v \bmod 2$, s is a diagonal in H.

The theorem thus distinguishes two cases according to the parity of the difference $u - v$, that is, according to whether the difference is odd or even.

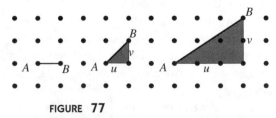
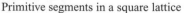

FIGURE 77

Primitive segments in a square lattice

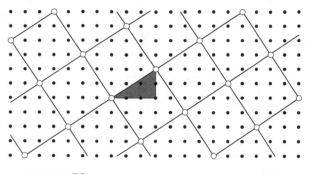

FIGURE 78
The primitive segment defines a new square lattice.

Figure 79 shows the example $u = 3$, $v = 2$ with $u \not\equiv v \bmod 2$, and $e = \sqrt{13}$. The axis of reflection s is a lattice-line in H.

Figure 80 shows the example $u = 3$, $v = 1$ with $u = v \bmod 2$ and $e = \frac{\sqrt{10}}{\sqrt{2}} = \sqrt{5}$. The axis of reflection s is a diagonal in H.

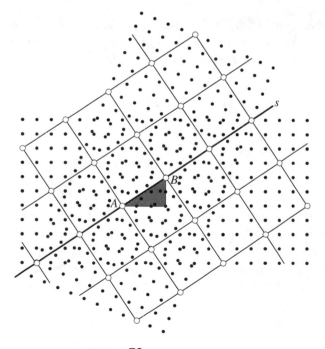

FIGURE 79
The example $u = 3$ and $v = 2$

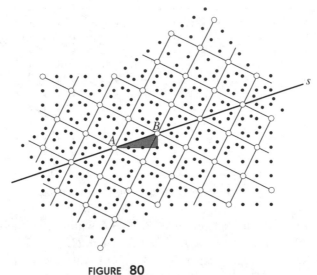

FIGURE **80**
The example $u = 3$ and $v = 1$

In the example of Figure 79 ($u \not\equiv v \bmod 2$), H is the square lattice determined by the primitive segment $\overline{AB}$; in the example in Figure 80 ($u \equiv v \bmod 2$) the midpoints of the squares also belong to H.

4.6 PYTHAGOREAN TRIANGLES

Pythagorean triangles, that is, right-angled triangles with integer side-lengths, and the associated number triples ($a, b, c \in \mathbb{N}, \quad a^2 + b^2 = c^2$) are mostly considered in their number-theoretical aspects, wherein, in particular, questions of divisibility play a role. [15] gives a broad introduction to this order of ideas, from an instructional point of view, and contains an ample bibliography.

Pythagorean triangles have also, however, a visual aspect, if square lattices and their superimpositions are used. As an example, let us draw, in a square lattice G, the well-known Pythagorean triangle with cathetus-lengths 3 and 4 and hypotenuse-length 5, and let us subdivide the hypotenuse-square into its 25 unit squares (Figure 81a).

It is clear that the lattice G' in the hypotenuse-square has, in addition to the vertices of the square, further lattice points in common with the

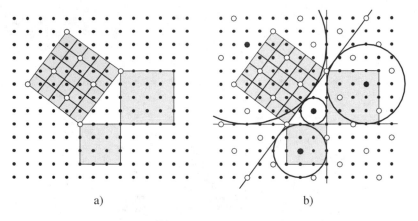

a)   b)

FIGURE 81
Square lattice in the hypotenuse-square

cathetus-lattice G. These common lattice points form a new square lattice, the covering lattice H with edge-length $\sqrt{5}$. We can interpret the square lattices G and G' as the sets of their lattice points and denote by $H := G \cap G'$ the set of points they have in common. Investigations of examples of other primitive Pythagorean triangles, that is, Pythagorean triangles with $\gcd(a, b, c) = 1$, lead to the following conjectures:

a) The cathetus-lattice G and the hypotenuse-lattice G' have as common lattice points a covering square lattice H of edge-length $\sqrt{c}$.

b) The centers of the inscribed circle and the escribed circles are lattice points of the covering lattice H (Figure 81b). From this it follows that the radii are whole numbers [3].

The proof of these conjectures is to be found in [49].

4.7 PARAMETRIZING THE PRIMITIVE TRIANGLES

The primitive Pythagorean triangles may be generated as follows: For $u, v \in \mathbb{N}$, with $u > v$, u, v coprime, and $u \neq v \bmod 2$, the quantities $a = u^2 - v^2$, $b = 2uv$, $c = u^2 + v^2$ are the sides of a primitive Pythagorean triangle. Conversely, to each primitive Pythagorean triangle with sides a, b, c there are unique integers u, v satisfying the above conditions [38].

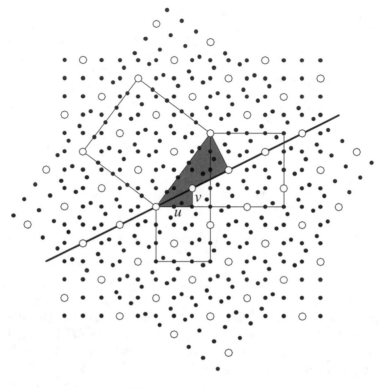

FIGURE **82**
The support triangle with cathetus u and v

These parameters u and v have a direct geometric significance. In the example $u = 2$ and $v = 1$, we obtain $a = 3$, $b = 4$ and $c = 5$. The parameters u and v are, in fact, the cathetus of the support triangle which determines the covering lattice H (Figure 82). The lattice line of H defined by this support triangle is an angle-bisector of the original right-angled triangle; the hypotenuse-lattice G' arises by reflecting the cathetus-lattice in this angle-bisector. If now a right-angled triangle with cathetus-lengths $u\sqrt{c}$, $v\sqrt{c}$ is drawn in the covering-lattice H, this becomes the support triangle of the hypotenuse c of the primitive Pythagorean triangle.

These properties of the covering lattice, here discussed by example, are valid in general for primitive Pythagorean triangles.

Question 4.13 Figure 83 shows a square lattice and its image under a $45°$ turn. Have the two lattices any point in common apart from the pivot?

FIGURE 83
Square lattices turned through 45°

Question 4.14 Through what angle β can a square lattice be turned so that it and its image have points in common apart from the pivot?

4.8 IN A REGULAR TRIANGULAR LATTICE

We study now triangles with integer side-lengths, and with two sides along lattice lines in a regular triangular lattice G. These two sides, which we denote by a and b by analogy with the cathetus of a right-angled triangle, enclose at the vertex C an angle of 60° or 120°. For the side c we obtain from the cosine rule, for $\gamma = 60°$: $c^2 = a^2 + b^2 - ab$, and for $\gamma = 120°$: $c^2 = a^2 + b^2 + ab$. An example of the second case is $a = 3$, $b = 5$, and $c = 7$ (Figure 84a).

From such a triangle, with $\gamma = 120°$, a triangle with $\gamma = 60°$ can be constructed by adjoining an equilateral triangle, in two different ways (Figure 84b). We restrict ourselves in what follows to integer triangles

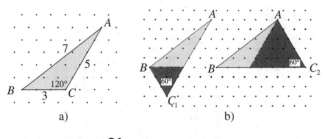

a)   b)

FIGURE 84

Integer triangles in the triangular lattice

with $\gamma = 120°$ and $\gcd(a, b, c) = 1$. Following [11, p. 405] and [19], these primitive 120°-triangles can be parametrized as follows: With $u, v \in \mathbb{N}$, $u > v$, u, v coprime, $u \not\equiv v \bmod 3$, we set $a = u^2 - v^2$, $b = 2uv + v^2$, $c = u^2 + v^2 + uv$. For $u = 2$ and $v = 1$ we get the example $a = 3$, $b = 5$ and $c = 7$. In analogy with the primitive Pythagorean triangles, we can reflect the lattice G by the angle bisector w_α in the side c. This reflected lattice is called G'. It may be shown that $H := G \cap G'$ is an equilateral-triangle lattice (Figure 85).

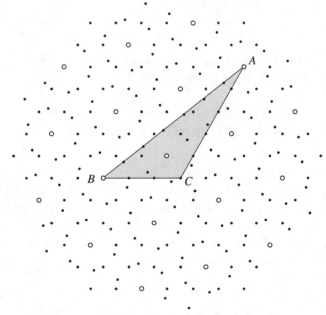

FIGURE 85

In the triangular lattice

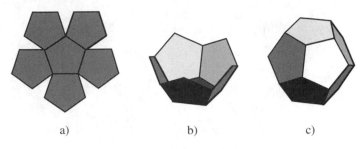

a) b) c)

FIGURE 86
Regular pentagons

Answers to Questions

Answer 4.1 In order that a parquet piece fits at a vertex, it is necessary that the interior angle be a factor of 360°. This is the case only with equilateral triangles, squares and regular hexagons. With a regular pentagon, for example, with the interior angle of 108°, there remains a gap of 36° (Figure 86a). This gap can be eliminated by bending up the adjacent pentagons; we thereby leave the plane and obtain a cup in space (Figure 86b). If we put a second cup on as a roof, we obtain a pentagonal dodecahedron (Figure 86c). The pentagonal dodecahedrom consists of 12 regular pentagons; 3 pentagons come together at each vertex.

Answer 4.2 Figure 87 shows two possibilities. The parquet of Figure 87a is an affinely skewed version of the chessboard parquet; the parquet pieces all

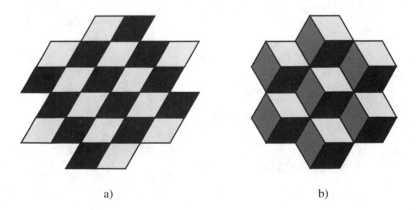

a) b)

FIGURE 87
Parqueting with 60°-rhombi

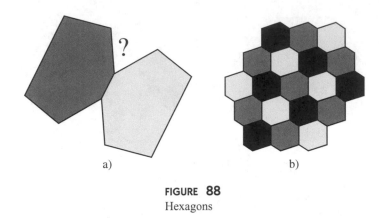

a)   b)

FIGURE 88
Hexagons

have the same direction, and a "displacement", that is, a shift of part of the parquet by an arbitrary multiple of the edge length, is always possible. In the parquet of Figure 87b we have tiles in 3 different directions; no displacement is possible. This parquet creates an illusion: It takes effort to see the parquet as "flat" and not as a collection of cubes.

Answer 4.3 No parquet can be made with an arbitrary hexagon, because the angles coming together at a vertex do not add up to a full rotation (Figure 88a). On the other hand, a parquet can be laid out with point-symmetric hexagons—on a point-symmetric hexagon opposite sides are parallel (Figure 88b).

Answer 4.4 Equilateral triangles and regular hexagons can be combined (Figure 89a) as can equilateral triangles, squares and regular hexagons (Figure 89b). On the other hand, it is not possible to combine only squares

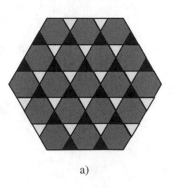
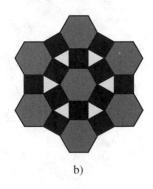

a)   b)

FIGURE 89
Parquets with regular polygons

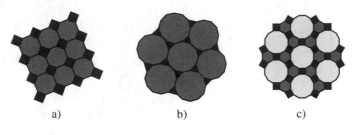

a)   b)   c)

FIGURE 90
Further possibilities

and regular hexagons. But there are other possible combinations, if we also admit regular octagons and dodecagons (Figure 90). There are, in fact, many other (non-regular) possibilities.

Answer 4.5 The highlighted "propeller" of Figure 91a has a 3-fold rotational symmetry, that is, under a one-third turn—i.e., a rotation through 120°—it returns to its original position. Thus the four centers of the triangles belonging to the propeller form angles of 120°. We reason similarly for all other propeller points and conclude that the centers of the triangles form equilateral triangles (compare [20, p. 24 ff]).

If we only keep in view one of the general triangles and its three adjacent equilateral triangles (Figure 91b), then the centers of the equilateral triangles form a new equilateral triangle. This situation is described as *Napoleon's Theorem;* it is said that it was discovered by Napoleon. Of course, this theorem can be proved directly, without recourse to parquets (compare [10, p. 68 ff]).

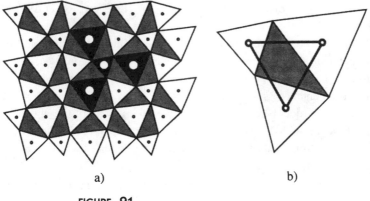

a)   b)

FIGURE 91
a) The propeller   b) Napoleon's Theorem

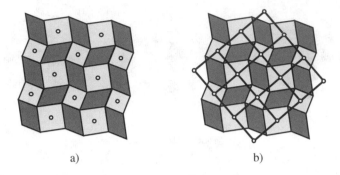

a)   b)

FIGURE 92
Parallelograms and squares

Answer 4.6 We obtain a parquet in which the centers of the squares themselves lie on a square lattice (Figure 92). The proof proceeds analogously to that of Napoleon's Theorem. The case with the parallelogram and the squares is a special case of the ***Theorem of Napoleon-Barlotti:*** *If regular n-gons are erected exterior to the sides of an affinely-regular n-gon, then their centers again form a regular n-gon.* This generalization cannot be proved by the use of parquets. Literature on Napoleon's Theorem and the more general Theorem of Napoleon-Barlotti may be found in [1], [8], [10], [25], [27], [42].

Answer 4.7 In theory, the Dirichlet parquet associated with the set of schoolhouses, regarded as a point-set, solves the problem. In practice, the likely traffic-effects must be taken into consideration.

Answer 4.8 With the regular hexagonal parquet, the associated Dirichlet parquet consists of equilateral triangles; its Dirichlet parquet is once more the regular hexagonal parquet (Figure 93). There are, of course, other examples.

Figure 94 shows three examples of parquets which are congruent to their Dirichlet parquets. In each, the Dirichlet parquet is orthogonal to the original parquet. These examples are derived from each other by turning the thickly drawn segmental crosses.

If such a parquet is put together using only one type of tile, this must be a quadrilateral with a circumcircle, that is, a so-called ***cyclic-quadrilateral.*** This follows from the fact that the right bisectors of the sides meet in a point. Figure 95 shows an example.

Answer 4.9 We need a twist-and-stretch transformation with an angle of rotation β and stretch-factor $\frac{c}{a} = \frac{1}{\cos \beta}$.

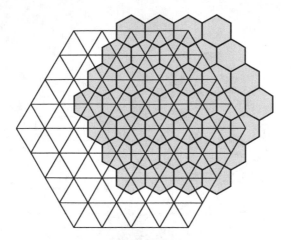

FIGURE 93
The regular hexagonal parquet

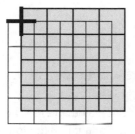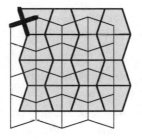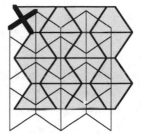

FIGURE 94
Congruence to the Dirichlet parquet

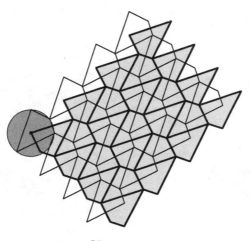

FIGURE 95
Parquet of cyclic-quadrilaterals

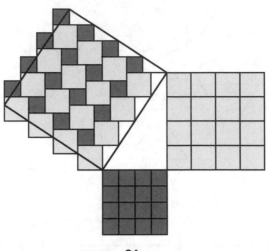

FIGURE 96
Parquet with squares

Answer 4.10 Figure 96 shows a parqueting with 16 squares. Such figures can also occur as a pattern of weaves (compare [17, p. 291 ff.]).

Answer 4.11 Figure 97a shows an example of the cathetus theorem. The triangle on the hypotenuse must be subdivided by a line running to the foot of the altitude of the right-angled triangle. Figure 97b illustrates Pythagoras' Theorem with parquets. In the second cathetus-triangle there occurs a regular hexagonal parqueting, which has a too large total surface area. To solve the problem correctly, the cathetus-triangles should be replaced by cathetus-hexagons (compare Figure 73).

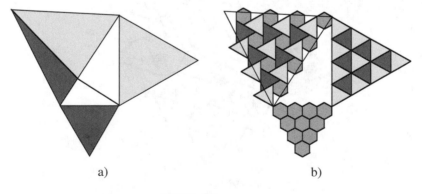

a)   b)

FIGURE 97
Equilateral triangles

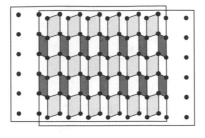

FIGURE 98
Parallel coverings

Answer 4.12 There results a grid with two different parallelograms (Figure 98).

Answer 4.13 The two grids have just the origin of rotation O in common. This can be seen indirectly: We assume there is a further common point P; this point P has, in the first grid, integer coordinates (p, q) relative to O, and in the second grid, at a $45°$ angle to the first, the integer coordinates (r, s) (Figure 99). Hence

$$r = \frac{1}{\sqrt{2}} p + \frac{1}{\sqrt{2}} q$$

$$s = -\frac{1}{\sqrt{2}} p + \frac{1}{\sqrt{2}} q.$$

From the first equation we get $\sqrt{2} = \frac{p+q}{r}$; since $\sqrt{2}$ is irrational whereas $\frac{p+q}{r}$ is rational, we have a contradiction.

Answer 4.14 For a further point P with integer coordinates (p, q) and (r, s) relative to the two points, it follows (Figure 100) that

$$r = p \cos \beta + q \sin \beta$$

$$s = -p \sin \beta + q \cos \beta.$$

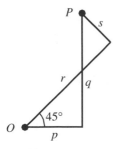

FIGURE 99
Two common points O and P

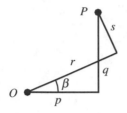

FIGURE 100

The general case

From this we obtain

$$\cos \beta = \frac{rp + sq}{p^2 + q^2}; \quad \sin \beta = \frac{rq - sp}{p^2 + q^2}.$$

Thus we obtain the necessary condition that $\cos \beta$ and $\sin \beta$ are both rational.

With a turn through 30° (Figure 101) there can therefore be no common point beyond the origin of rotation since $\cos 30°$ is irrational.

The condition that $\cos \beta$ and $\sin \beta$ are both rational is however also sufficient: In this case we can write them in the form $\cos \beta = \frac{a}{c}$, $\sin \beta = \frac{b}{c}$, with c the least common denominator. But then a, b, c are the sides of a

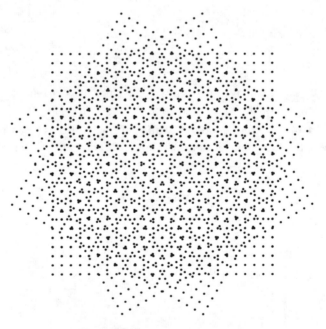

FIGURE 101

Square point-grids at an angle of 30°

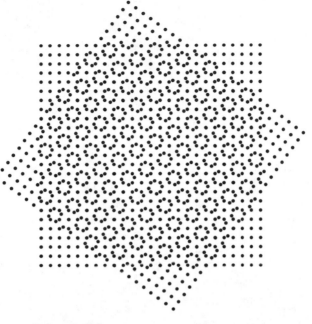

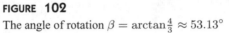

FIGURE 102

The angle of rotation $\beta = \arctan\frac{4}{3} \approx 53.13°$

primitive Pythagorean triangle with the angle β. The two point-grids thus have the two endpoints of the hypotenuse of this primitive Pythagorean triangle in common. Figure 102 shows the case of the angle

$$\beta = \arctan\frac{4}{3} \approx 53.13°,$$

so that $\cos\beta = \frac{3}{5}$ and $\sin\beta = \frac{4}{5}$.

The covering grid is identical with the covering grid of Figure 82.

CHAPTER 5

The Problem of the Center

5.1 WHERE IS THE CENTER OF THE WORLD?

In many pictures, figures and situations we speak quite naturally, and somewhat imprecisely, of a "center". This can be a point or a moment of time, or more rarely, a line or a plane. Figure 103 shows a few examples.

The question about the center also has an aesthetic aspect: What is the center (or center of gravity) of a picture, a piece of sculpture or a building (compare [2])?

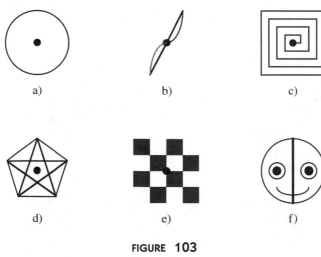

a)   b)   c)

d)   e)   f)

FIGURE 103
The center

Question 5.1   Is there a regular figure *without* a center?

Question 5.2   Are there also cases of *several* centers?

Question 5.3   How can the vertices of a square be optimally linked (to form a network of minimal length)?

A center can also be seen dynamically as the collision-point of two opposing motions. Let us give an example: In Figure 104 a closed curve arises from the repeated construction, by refraction, of circular arcs. (Such closures are described in [29], [47] and [48]).

Proof of the closure property can be carried out as follows. We denote by a_0, a_1, a_2 the 3 sides of the triangle $A_0 A_1 A_2$, and by r_0 the radius of the initial arc. For the radii of the subsequent circular arcs we then have:

$$r_1 = a_1 - r_0,$$
$$r_2 = a_2 - r_1 = a_2 - a_1 + r_0,$$
$$r_3 = a_0 - r_2 = a_0 - a_2 + a_1 - r_0,$$
$$r_4 = a_1 - r_3 = -a_0 + a_2 + r_0,$$
$$r_5 = a_2 - r_4 = a_0 - r_0,$$
$$r_6 = a_0 - r_5 = r_0.$$

From $r_6 = r_0$ the closure property follows. If now in Figure 104 we increase the initial radius r_0, then the radius r_2 is increased by the same amount, and the two points B_0 and B_3 retreat towards each other. Thus we have opposing movements of the two points. For the special initial radius $r_0 = \frac{a_0 + a_1 - a_2}{2}$ the two points B_0 and B_3 coincide, and we obtain a ***center line*** consisting of only 3 circular arcs (Figure 105a).

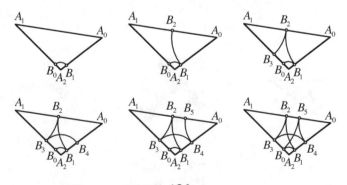

FIGURE 104

A figure that closes

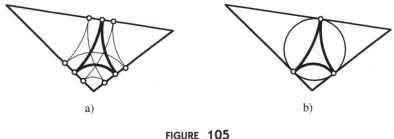

a)   b)

FIGURE 105
A center line

The vertices of this center line are precisely the points of contact of the incircle with the triangle (Figure 105b).

Question 5.4 What happens if we start at B_0 with a line parallel to the side A_0A_1 of the triangle and then continue as in Figure 106?

5.2 MEAN VALUES

5.2.1 Half is Eaten

An ice on a stick (Figure 107) is half-eaten at the $\frac{a+b}{2}$ mark, where a is the lower limit and b is the upper limit of the ice, always assuming that the ice is eaten exclusively from the top down. The mean-mark is calculated as the *average* or *arithmetic mean* $\frac{a+b}{2}$ of the two boundaries a and b of the ice. But there are many examples where the center cannot be identified with the arithmetic mean.

Question 5.5 When is a cylindrical ice, which is licked away all round in regular fashion, half consumed?

Question 5.6 When is a spherical candy half gone?

Question 5.7 When is a toilet roll half used up?

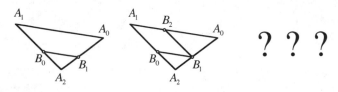

FIGURE 106
How does it continue?

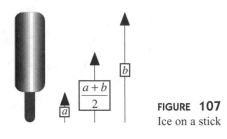

FIGURE 107
Ice on a stick

5.2.2 Average Speed

There are cases in which, in the same context, different methods of finding the mean must be employed according to the precise question put. A well-known example of this is the question about the average speed of a car, which first travels with a speed $v_1 = 60 \frac{\text{km}}{\text{hr}}$ and then with a higher speed of $v_2 = 80 \frac{\text{km}}{\text{hr}}$.

If the car travels for the *same length of time* t at each of the two speeds v_1 and v_2, we obtain the average speed

$$\bar{v} = \frac{v_1 t + v_2 t}{2t} = \frac{v_1 + v_2}{2}.$$

The average speed is thus the arithmetic mean of the two individual speeds. As the question was put, it is quite possible that one or both of the speeds is negative; the associated stretch of the course would then have been traversed backwards.

If, on the other hand, the car had traveled over the same distance s at the speeds v_1 and v_2, we would have obtained for the travel times of the two segments $t_1 = \frac{s}{v_1}$, $t_2 = \frac{s}{v_2}$ so that the average speed is

$$\tilde{v} = \frac{2s}{t_1 + t_2} = \frac{2s}{\frac{s}{v_1} + \frac{s}{v_2}} = \frac{2}{\frac{1}{v_1} + \frac{1}{v_2}} = \frac{2v_1 v_2}{v_1 + v_2}.$$

This is the so-called ***harmonic mean.*** In this way of putting the precise question, we could not allow v_1 or v_2 to be negative, because then the corresponding time-interval would have been negative. The harmonic mean can also be written in the form

$$\frac{1}{\tilde{v}} = \frac{1}{2}\left(\frac{1}{v_1} + \frac{1}{v_2}\right);$$

the reciprocal of the harmonic mean is thus the arithmetic mean of the reciprocals of the original quantities. In our numerical example we obtain for the arithmetic mean $\bar{v} = 70 \frac{\text{km}}{\text{hr}}$ and, for the harmonic mean, $\tilde{v} \approx 68.57 \frac{\text{km}}{\text{hr}}$.

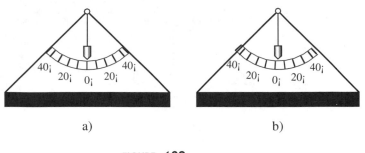

FIGURE 108
Defective angle-measure

Question 5.8 Is the harmonic mean always smaller than the arithmetic mean?

5.2.3 Correcting Systematic Errors

With every measurement an error arises. Especially to be feared are systematic errors, which, for example, may be due to a technical defect in the measuring apparatus or a consistently erroneous use of the apparatus.

A rather old-fashioned angle-measure (Figure 108a), which, with the help of a plumb-line, can measure the angle of inclination to the horizontal, can serve as an example. If the dial of the apparatus is somewhat skew (Figure 108b), a systematic error arises.

This systematic error σ can however very easily be eliminated. To determine an angle of inclination α, we make two measurements, turning the measuring apparatus through $180°$ between the measurements (Figure 109).

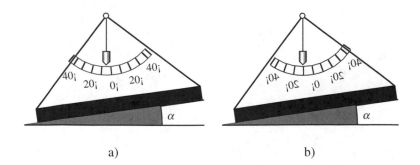

FIGURE 109
Error elimination

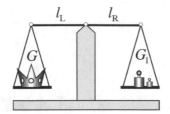

FIGURE 110
A balance scale

With the first measurement we get a reading of $\alpha_1 = \alpha - \sigma$ and, with the second measurement $\alpha_2 = \alpha + \sigma$. The actual value α is the arithmetic mean of the two readings.

In the next example it is a matter of measuring a weight G with a scale balance (compare [23]). To that end we place the object of weight G in the left-hand pan and put weights in the right-hand pan until the balance is in equilibrium (Figure 110). The sum G_1 of the weights is the outcome of the first measurement.

A systematic error can arise if the lengths ℓ_L and ℓ_R of the left and right lever arms of the balance are not equal. To eliminate this systematic error, we exchange the roles of the two pans. We place the object to be weighed in the right pan and balance it with weights in the left pan (Figure 111). In this way we obtain a second measurement G_2 for the weight.

How do we arrive at the true weight G from the two measurements G_1 and G_2? From the law of the lever we get, from the first weighing: $\ell_L G = \ell_R G_1$. Thus $G = \frac{\ell_R}{\ell_L} G_1$. From the second weighing we obtain,

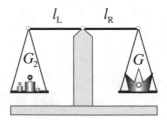

FIGURE 111
The second weighing

similarly, $\ell_L G_2 = \ell_R G,$ so $G = \frac{\ell_L}{\ell_R} G_2$. From this we obtain

$$G^2 = \frac{\ell_R}{\ell_L} G_1 \frac{\ell_L}{\ell_R} G_2 = G_1 G_2$$

and, finally, $G = \sqrt{G_1 G_2}$. G is the so-called ***geometric mean*** of G_1 and G_2.

Remark   In the example of the angle-measure, the systematic error $\pm\sigma$ is an *additive* error. An additive error can be eliminated by taking an arithmetic mean, because it is canceled by taking a sum. In the example of the scale balance, the systematic error is given by the *factor* $(\frac{\ell_R}{\ell_L})^{\pm 1}$. This error is eliminated by taking the product in the geometric mean.

Question 5.9   Would one obtain for G a value that is too large or too small if, instead of taking the geometric mean, one erroneously took the arithmetic mean?

The simplest example for the arithmetic and geometric means is provided by the "symmetrization" of a rectangle, that is, the replacement of a rectangle with sides of length a and b by a square of side s. Given the condition that the *perimeter* should be held constant, one obtains for s the arithmetic mean $s = \frac{a+b}{2}$. This is not surprising since the perimeter is an additive function of the lengths of the sides. But, given the condition that the *enclosed area* should remain constant, one obtains for s the geometric mean $s = \sqrt{ab}$, since the area is a multiplicative function of the lengths of the sides.

Question 5.10   What do we obtain when we symmetrize a rectangular parallelepiped with edges of length a, b, c?

5.2.4 Minimal Supply Channels

Along a street stand 5 houses (Figure 112); on the same street a central office Z is to be built, which will provide direct cable service to each house individually. What is the optimal position for the office?

FIGURE 112
Where should the central office be built?

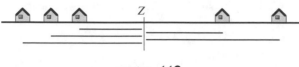

FIGURE 113
A trial

We find the solution immediately if we draw in, experimentally, the required cables to the individual houses from an arbitrarily chosen position for Z (Figure 113).

If we now move the point Z in Figure 113 to the left, 3 cables are shortened and 2 are lengthened, by the same amount—the total cable-length is thus shortened. This holds right until we bring Z into coincidence with the "middle" house—even if this house is not geometrically at the midpoint of the entire street.

Question 5.11 Where would we build the central office when we have to service 6 (more generally an even number of) houses?

The middle which occurs here is called the ***median***. The determination of the median presupposes that we can arrange the elements in a sequence. In our example this was the ordering of the houses along the street. Numbers may be ordered according to size. We find the median as follows. First we order the elements and strike out the first and last elements in the sequence. From the remaining elements we again strike out the first and last elements, and so on. The element surviving at the end, assuming there were originally an odd number of elements, is the median. Thus, for example, the numbers $6, 3, 61, 5, 12, 13, 5$ yield the process

$$3 < 5 = 5 < 6 < 12 < 13 < 65$$

$$5 = 5 < 6 < 12 < 13$$

$$5 < 6 < 12$$

$$6,$$

and so have the median 6, which is far removed from the arithmetic mean 15.

Answers to the Questions

Answer 5.1 The band-design of Figure 114, thought of as infinitely long, is translation-symmetric, but contains no "middle".

FIGURE 114
A translation-symmetric band-design

Answer 5.2 If we think of the chessboard pattern of Figure 103e as extended indefinitely, every vertex of a square and every midpoint of a black or white square can serve as a midpoint. In the same way, every point on a straight line, and every point in a plane, can be interpreted as a midpoint. The same may be said to be true also for arbitrary points on a circle or a sphere. Thus if some country on this earth sees itself as being at the "center" of the earth, that is not a geometrical or geographical statement, but a political statement. Figure 115 shows a band-design which, in addition to translation-symmetry, contains point-symmetries and thus infinitely many "midpoints".

Answer 5.3 The obvious idea of connecting the 4 vertices of the unit square with the midpoint of the square (Figure 116a) does not provide the best solution. For this linkage the total length is $2\sqrt{2} \approx 2.828$. If instead we choose two branch points on a mean parallel such that the three paths from each such point are at $120°$ angles to each other (Figure 116b), then the line segments to a vertex have length $\frac{\sqrt{3}}{3}$, while the distance between the branch points has length $\frac{3-\sqrt{3}}{3}$; thus the entire network of paths has length $1 + \sqrt{3} \approx 2.732$. In fact, this is optimal—but, of course, is not unique: rotation by $90°$ yields a different network.

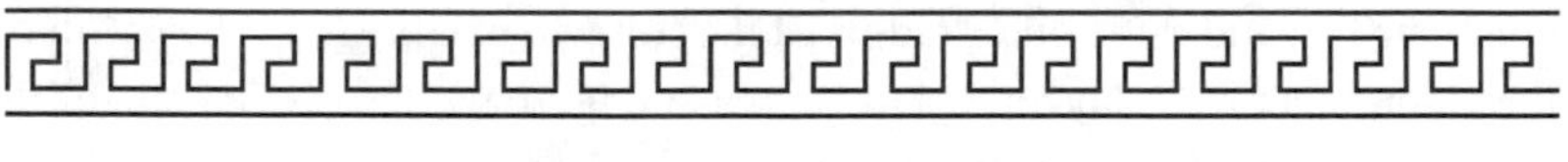

FIGURE 115
Band-design with point-symmetries

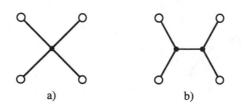

a)   b)

FIGURE 116
Two networks of paths in the square

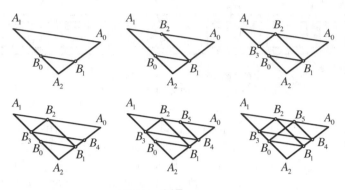

FIGURE **117**
A figure which closes

Answer 5.4 A closed figure arises in accordance with Figure 117. The proof of the closure property consists of transferring the proportionality relations of the points B_i with respect to the appropriate sides of the triangle by means of the proportionality theorems. Here, too, there is a mean line: its vertices are the midpoints of the sides of the triangle (Figure 118).

Answer 5.5 From the formula $\pi r^2 h$ for the volume of a cylinder it follows that a cylinder of the same height h but only half the volume requires a radius $\frac{1}{\sqrt{2}}r \approx 0.707r$. The ice is therefore half eaten when there is roughly 70% of the diameter remaining. Here the thickness of the stick has been ignored.

Answer 5.6 From the formula $\frac{4}{3}\pi r^3$ for the volume of a sphere it follows that a sphere of half the volume will have a radius of $\frac{1}{\sqrt[3]{2}}r \approx 0.794r$. Thus half the volume is gone when around 80% of the diameter remains.

Answer 5.7 Let r be the inner radius of the roll, R the outer radius of the complete roll and x the outer radius of the roll when half used. The width of the toilet-paper will play no part in our reasoning, so we can restrict ourselves to a calculation of cross-sections. From

$$\pi x^2 - \pi r^2 = \frac{1}{2}(\pi R^2 - \pi r^2)$$

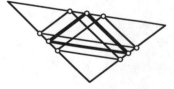

FIGURE **118**
Mean line

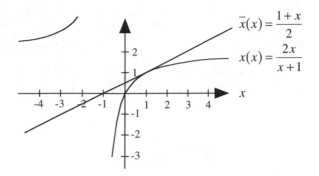

$$\bar{x}(x) = \frac{1+x}{2}$$

$$x(x) = \frac{2x}{x+1}$$

FIGURE 119
The graphs of the two functions

we conclude that the "toilet-roll mean" is given by

$$x = \sqrt{\frac{R^2 + r^2}{2}}.$$

Answer 5.8 In the case $v_1 = v_2$ the arithmetic and harmonic means are equal. If $v_1 \neq v_2$, we will choose the measurement units so that v_1 has measure 1. The two means now depend only on the measure x of v_2. Since the question postulates a harmonic mean we must have $v_2 > 0$, so $x > 0$. For the arithmetic mean $\bar{x}$ we have the linear relation $\bar{x}(x) = \frac{1+x}{2}$, and for the harmonic mean $\widetilde{x}$ the relation $\widetilde{x}(x) = \frac{2x}{x+1}$. Figure 119 shows the graphs of the two functions. For $x > 0$ we thus have $\bar{x} \geq \widetilde{x}$ with equality precisely when $x = 1$.

Answer 5.9 The arithmetic mean of two distinct numbers p and q is bigger than the geometric mean of these numbers. To see this we interpret p and q as segments of the hypotenuse of a right-angled triangle (Figure 120). The arithmetic mean $\frac{p+q}{2}$ is half the length of the hypotenuse, thus the radius of the circle of Thales. By a well-known altitude theorem the height $h = \sqrt{pq}$ is the geometric mean of p and q. For $p \neq q$ the height h is smaller than the radius of the circle of Thales. Notice that in Figure 120b, $\overline{AD} = q$, $\overline{DB} = p$.

The harmonic mean of p and q is $\frac{2pq}{p+q}$. Since $\frac{p+q}{2} \frac{2pq}{p+q} = pq = h^2$, it follows from the cathetus theorem for the right-angled triangle CMD (Figure 120b), that the harmonic mean is the hypotenuse segment $\overline{CE}$. This is smaller than the cathetus h; but that means that the harmonic mean is smaller than the geometric mean.

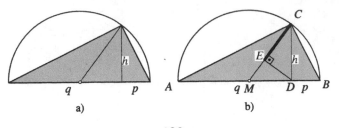

a)   b)

FIGURE 120

Hypotenuse sections

We could also argue in this way. If A, G, H are the arithmetic, geometric and harmonic means, then $G < A$ and $AH = G^2$. Thus $H = \frac{G}{A}G < G$.

Answer 5.10 We change the rectangular parallelepiped into a cube of edge-length s. Under the condition that the total sum of the edge-lengths remains the same, we obtain for s the generalized arithmetic mean $s = \frac{a+b+c}{3}$. If the surface area is to remain the same, we obtain $s = \sqrt{\frac{ab+bc+ca}{3}}$. This is a mixture of arithmetic and geometric means. Finally, if the volume is to remain unchanged, we require the generalized geometric mean $s = \sqrt[3]{abc}$.

Answer 5.11 At the end of the striking-out process there remain, in the middle, two houses left over. The central office may be built anywhere on the stretch between these two houses.

CHAPTER 6

Symmetry in Word, Script and Number

[Translator's note: Much of the material of this short chapter, in the original German version, depends on a deep appreciation of the German language. With the author's approval, this material has been omitted.]

6.1 PALINDROMES

By a *palindrome* we understand a meaningful sequence of letters or words which, when read backwards, produces the same or perhaps a different meaning. The term *palindrome* is derived from the Greek palindromos, "running backwards". Examples are the proper names (compare [44]):

ANNA

OTTO

ANNA SUSANNA

The proper name "OTTO" has also a typographical axial symmetry with a vertical axis of symmetry in the middle of the word.

The palindromic sentence

ABLE WAS I ERE I SAW ELBA

has been—ironically—attributed to Napoleon. In token of the increasing scarcity of open-handed generosity, perhaps some day soon the slogan

SEX AT NOON TAXES

will appear. A palindrome which has enjoyed some popularity in the United States is

A MAN, A PLAN, A CANAL—PANAMA.

The translator invented, in 1945, the pleasing palindrome

DOC, NOTE, I DISSENT. A FAST NEVER PREVENTS A FATNESS. I DIET ON COD.

Try making up your own. If you don't succeed you'll find many on the web at http://www.cosy.sbg.ac.at/~leo/palindrom/

6.2 PALINDROMIC NUMBERS

Palindromic numbers have a symmetric ordering of their digits, for example, 77, 252, 3443. It then transpires that every palindromic number with an *even* number of digits is divisible by 11 (compare [21, p. 37]).

Question 6.1 How could this be proved?

Question 6.2 Is there a corresponding theorem for palindromic numbers in base p?

If we add to an arbitrary natural number its mirror-number, that is, the number we obtain by reversing the digits, and if we repeat this process indefinitely, we obtain, as a general rule, a palindromic number. For example, with the starting number 1944 we obtain

$$\begin{array}{lll} \text{Step 1:} & 1944 + 4491 = 6435 \\ \text{Step 2:} & 6435 + 5346 = 11781 \\ \text{Step 3:} & 11781 + 18711 = 30492 \\ \text{Step 4:} & 30492 + 29403 = 59895 \end{array}$$

The question, whether each starting number leads in a finite number of steps to a palindromic number, remains open. There are numbers, for example the number 196, of which it is conjectured that they never lead to a palindromic number [28], [30].

6.3 RHYMING SCHEMES

Figure 121 shows different rhyming schemes in two consecutive four-line verses from *Ivanhoe,* by Sir Walter Scott (Penguin Books):

THE BARE FOOTED FRIAR
I give thee, good fellow, a twelvemonth or twain,
To search Europe through, from Byzantium to Spain;
But ne'er shall you find, should you search till you tire,
So happy a man as the Barefooted Friar.

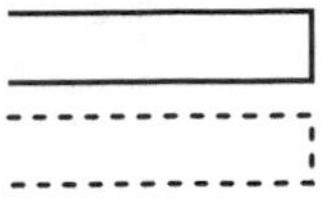

KNIGHT AND WAMBA
There came three merry men from south, west, and north
    Ever more sing the roundelay;
To win the widow of Wycombe forth,
    And where was the widow might say them nay?

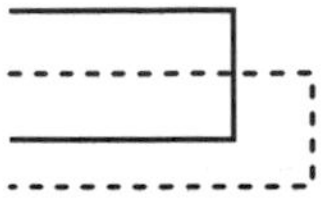

FIGURE **121**
Two different rhyming schemes.

Question 6.3 How many different rhyming schemes are there for four-line verses?

We study now n-line verses with k different line-endings. Let $S(n, k)$ be the number of associated rhyming schemes. For one-line verses there is only one scheme, so $S(1, 1) = 1$. For $S(n + 1, k)$ we obtain a recursion formula by separating the possibilities for the $(n + 1)$st line into two cases:

1. The last line, that is, the $(n + 1)$st line, has its own unique ending. Then there remain for the previous n lines $(k - 1)$ endings available, so that there are $S(n, k - 1)$ possibilities.

2. The $(n + 1)$st line has one of the k already used endings. In this case there are then $kS(n, k)$ possibilities.

Putting these together, we obtain:

$$S(n + 1, k) = S(n, k - 1) + kS(n, k).$$

This recursion formula is, except for the factor k, just like that for binomial coefficients in *Pascal's Triangle.* (It is implicit that $S(n, 0) = 0$ and $S(n, k) = 0$ if $n < k$.] With the starting value $S(1, 1) = 1$ one obtains from the recursion-formula the triangle of so-called *Stirling numbers of the second kind.* (With the Stirling numbers one begins counting at 1, in contrast with the binomial coefficients.) In Figure 122 the recursion formula is indicated by the system of arrows. Stirling numbers of the first and second kind play an important role in combinatorics ([9], [24], [46]).

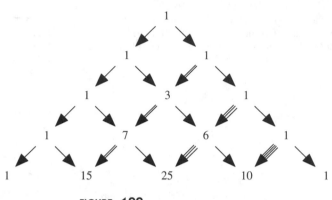

FIGURE 122
Stirling numbers of the second kind

For the total number of rhyming schemes for an n-line verse, one must add up all the Stirling numbers of the second kind in the nth row. This leads to the **Bell numbers** $B(n) = \sum_{k=1}^{n} S(n, k)$.

Answers to Questions

Answer 6.1 For the number 11 we have, in the decimal system, the following divisibility rule: A number is divisible by 11 if and only if the alternating sum of its digits is divisible by 11. This is obviously the case for palindromic numbers with an even number of digits, since, for them, the alternating sum of the digits is zero.

Answer 6.2 A palindromic number in base p, with an even number of digits, is divisible by $(p + 1)$. So, for example, in base 2, the numbers $11(= 3)$, $1001(= 9)$, $1111(= 15)$, $100001(= 33)$, $101101(= 45)$, $110011(= 51)$, $111111(= 63)$, $\cdots$ are divisible by 3. In fact, for numbers written in base p, the divisibility rule corresponding to the rule-of-eleven above asserts that a number is divisible by $(p + 1)$ if and only if the alternating sum of its digits is divisible by $(p + 1)$. This can be seen as follows: from

$$a^n - b^n = (a - b)(a^{n-1} + a^{n-2}b + a^{n-3}b^2 + \cdots + b^{n-1})$$

it follows that $(a - b)$ is a factor of $a^n - b^n$, that is, $(a - b)|(a^n - b^n)$. With $a = p, b = -1$, this becomes

$$(p + 1)|\,(p^n - (-1)^n)$$

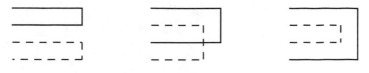

FIGURE 123
Rhyming schemes with two times two rhymes

and so $p^n \equiv (-1)^n \mathrm{mod}\,(p + 1)$. For a number z with digits z_k, $k \in \{0, 1, 2, \ldots, s\}$ in base p, that is, $z = \sum_{k=0}^{s} z_k p^k$, we thus have

$$z \equiv \sum_{k=0}^{s} z_k (-1)^k \mathrm{mod}\,(p + 1).$$

This shows that z is divisible by $(p + 1)$ exactly when $\sum_{k=0}^{s} z_k (-1)^k$ is divisible by $(p + 1)$.

Answer 6.3 If we assume that the 4 lines must be divided into 2 pairs of rhyming lines, there are just 3 schemes, as in Figure 123. If arbitrary rhymes are allowed, there are the 15 schemes of Figure 124. We have, trivially, one case with just one rhyme, that is, with 4 identical line-endings; then 7 cases with two line-endings; 6 cases with three line-endings; and finally the case without rhymes, that is, with four different line-endings (compare [46, p. 162]).

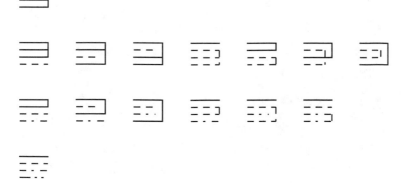

FIGURE 124
Rhyming schemes for 4-line verses.

References

[1] Alpers, K., Zu den Sätzen von Pythagoras und Napoleon über Symmetriebetrachtungen. *Mathematik in der Schule 35,* 1997, Heft 12, pp. 654–662.

[2] Arnheim, R., *Die Macht der Mitte. Eine Kompositionslehre für die bildenden Künste.* Köln: DuMont, 1983.

[3] Baptist, P., Inkreisradius und pythagoreische Zahlentripel. *Praxis der Mathematik 24,* 1982, pp. 161–164.

[4] Binninger, S., Die Fastspiegelung. Eine nichtaffine Verbindung von Achsenspiegelung und Punktspiegelung. *Praxis der Mathematik 38,* 1996, pp. 245–249.

[5] Bongartz, K., W. Borho, D. Mertens, und A. Steins, *Farbige Parkette. Mathematische Theorie und Ausführung mit dem Computer. Vier Aufsätze zur ebenen Kristallographie.* Basel, Boston, Berlin: Birkhäuser-Verlag, 1988.

[6] Bigalke, H.-G. und H. Wippermann, *Reguläre Parkettierungen. Mit Anwendungen in Kristallographie, Industrie, Baugewerbe, Design und Kunst.* Mannheim, Leipzig, Wien, Zürich: BI-Wissenschaftsverlag, 1994.

[7] Bourgoin, J., *Arabic Geometrical Pattern and Design.* New York: Dover, 1973.

[8] Christl, M., Der Satz von Napoleon im Schulunterricht. *Didaktik der Mathematik 22,* 1994, pp. 196–216.

[9] Cohen, D. I. A., *Basic Techniques of Combinatorial Theory*. New York: Wiley, 1978.

[10] Coxeter, H. S. M. and S. L. Greitzer, *Geometry Revisited*. Washington: The Mathematical Association of America, 1967.

[11] Dickson, L. E., *History of the Theory of Numbers, II. Diophantine Analysis*. Washington: Carnegie Institution, 1920.

[12] Ernst, B., *Der Zauberspiegel des Maurits Cornelis Escher*. Berlin: Taco, 1986.

[13] Escher, M. C., *Graphik und Zeichnungen*. München: Moos, 1984.

[14] Flachsmeyer, J., U. Feiste, und K. Manteuffel, *Mathematik und ornamentale Kunstformen*. Leipzig: Teubner-Verlag, 1990.

[15] Fraedrich, A. M., Pythagoreische Zahlentripel: Unterrichtliche Zugänge, Konstruktionsverfahren, sich anschließende Probleme und weiterführende Fragestellungen. *Didaktik der Mathematik 13,* 1985, pp. 31–49 und *Didaktik der Mathematik 13,* 1985, pp. 98–117.

[16] Fraedrich, A. M., *Die Satzgruppe des Pythagoras*. Mannheim, Leipzig, Wien, Zürich: BI-Wissenschaftsverlag, 1995.

[17] Gerdes, P., *Ethnogeometrie. Kulturanthropologische Beiträge zur Genese und Didaktik der Geometrie*. Bad Salzdetfurth: Franzbecker, 1990.

[18] Grünbaum, B. and G. C. Shephard, *Tilings and Patterns*. New York: Freeman, 1987.

[19] Hasse, H., Ein Analogon zu den ganzzahligen pythagoräischen Dreiecken. *Elemente der Mathematik 32,* 1977, pp. 1–6.

[20] Heilbronner, E. and J. D. Dunitz, *Reflections on Symmetry in Chemistry. . . . and Elsewhere*. Basel: Verlag Helvetica Chimica Acta, 1993.

[21] Hilton, P., D. Holton, and J. Pedersen, *Mathematical Reflections: In a Room with Many Mirrors, Second printing*. New York: Springer-Verlag, 1998.

[22] Hohler, F., *Mani Matter. Ein Porträtband. 2. Aufl.* Zürich: Benziger-Verlag, 1992.

[23] Humenberger, H., Exaktifizieren im Mathematikunterricht – am Beispiel des Begriffes „besser". *Der Mathematikunterricht,* 1996, pp. 71–79.

[24] Jeger, M., *Einführung in die Kombinatorik. Band 2.* Stuttgart: Klett-Verlag, 1976.

[24a] Johnson, Roger A., *Advanced Euclidean geometry: An elementary treatise on the geometry of the triangle and the circle.* Under the editorship of John Wesley Young, Dover Publications, Inc., New York, 1960.

[25] Kahle, D., Eine Bemerkung zum Satz von Napoleon-Barlotti für das Parallelogramm. *Didaktik der Mathematik 22,* 1994, pp. 217–218.

[26] Kirsch, A., Bemerkungen zum Vierecksschwerpunkt. *Didaktik der Mathematik 15,* 1987, pp. 34–36.

[27] Kratz, J., Vom regulären Fünfeck zum Satz von Napoleon-Barlotti. *Didaktik der Mathematik 20,* 1992, pp. 261–270.

[28] Kröber, G., Über Ergebnistypen und Muster in Palindromisierungs-prozessen. *Die $\sqrt{Wurzel}$ 3+4,* 1995, pp. 50–53.

[29] Kroll, W., Rundwege und Kreuzfahrten. *PM Praxis der Mathematik 32,* 1990, pp. 1–9.

[30] Lehmer, D., Sujets d'étude. *Sphinx 8,* 1938, pp. 12–13.

[31] Lindgren, H., *Geometric Dissections.* Revised and enlarged by Greg Frederickson. New York: Dover, 1972.

[32] Locher, J. C. (Herausgeber), *Leben und Werk M. C. Eschers.* Mit dem Gesamtverzeichnis des Graphischen Werks. Eltville am Rhein: Rheingauer Verlagsgesellschaft, 1986.

[33] Mandelbrot, B. B., *Die fraktale Geometrie der Natur.* Basel: Birkhäuser-Verlag 1991.

[34] Mandelbrot, B. B., *The Fractal Geometry of Nature.* New York: Freeman, 1983.

[35] Matter, M., *Us emene lääre Gygechaschte. Berndeutsche Chansons. 25. Aufl.* Zürich: Benziger-Verlag, 1993.

[36] Mazzola, G. (Herausgeber), *Symmetrie in Kunst, Natur und Wissenschaft. Ausstellungskatalog der gleichnamigen Ausstellung auf der Mathildenhöhe Darmstadt, 1. Juni bis 24. August 1986.* Darmstadt: Roether, 1986.

[37] Pfeiffer, H., *Oh Cello voll Echo. Palindromgedichte.* Frankfurt a. M.: Insel-Verlag, 1992.

[38] Rademacher, H. and O. Toeplitz, *Von Zahlen und Figuren*. Berlin, Heidelberg, New York: Springer-Verlag, 1968.

[38a] Rigby, John F., Orthocentric and mediocentric pentagons. *J. Geom.* 51 (1994), no. 1–2, pp. 116–137.

[38b] ———, Tritangent centres, Pascal's theorem and Thébault's problem. *J. Geom.* 54 (1995), no. 1–2, pp. 134–147.

[38c] ———, Precise colourings of regular triangular tilings. *Math. Intelligencer* 20 (1998), no. 1, pp. 4–11.

[39] Rilke, R. M., *Duineser Elegien. Die Sonette an Orpheus.* Zürich: Manesse-Verlag, 1951.

[40] Rosen, J., *Symmetry Discovered.* Cambridge University Press, 1975.

[41] Schattschneider, D. and W. Walker, *M. C. Escher Kaleidocycles.* Tarquin Publications, Stradbroke, Diss, Norfolk, 1991.

[42] Schmidt, F., 200 Jahre französische Revolution. Problem und Satz von Napoleon mit Varianten. *Didaktik der Mathematik 18,* 1990, pp. 15–29.

[43] Seebach, K., Über Schwerpunkte von Dreiecken, Vierecken und Tetraedern. Teil 1: *Didaktik der Mathematik 11,* 1983, pp. 270–282. Teil 2: *Didaktik der Mathematik 12,* 1984, pp. 36–44.

[44] Stengel, H. G., *Anna Susanna. Ein Pendelbuch für Rechts- und Linksleser. 3. Aufl.* München: List-Verlag, 1996.

[45] Stewart, I. and M. Golubitsky, *Fearful Symmetry: Is God a Geometer?* Oxford: Blackwell, 1992.

[46] Walser, H., Stirlingsche Zahlen im Unterricht. *Didaktik der Mathematik 13,* 1985, pp. 150–168.

[47] ———, Schließungsfiguren. *Didaktik der Mathematik 19,* 1991, pp. 187–206.

[48] ———, Geometrische Schließungsfiguren im Unterricht. *PM Praxis der Mathematik 35,* 1993, pp. 77–84.

[49] ———, Pythagoreische Dreiecke in der Gittergeometrie. *Didaktik der Mathematik 23,* 1995, pp. 193–205.

[50] ———, *Der Goldene Schnitt. 2. Aufl.* Leipzig: Teubner-Verlag, 1996.

[51] Wille, R. (Herausgeber), *Symmetrie in Geistes- und Naturwis-senschaft*. Hauptvorträge und Diskussionen des Symmetrie Symposions an der Technischen Hochschule Darmstadt vom 13. bis 17. Juni 1986 im Rahmen des Symmetrieprojektes der Stadt Darmstadt. Berlin, Heidelberg, New York, London, Paris, Tokyo: Springer-Verlag, 1988.

[52] Weyl, H., *Symmetry*, Princeton University Press, 1983.

Index